Peyman Vaziri

Introduction to Renewable Energy Applications

Peyman Vaziri

Introduction to Renewable Energy Applications

Noor Publishing

Imprint

Any brand names and product names mentioned in this book are subject to trademark, brand or patent protection and are trademarks or registered trademarks of their respective holders. The use of brand names, product names, common names, trade names, product descriptions etc. even without a particular marking in this work is in no way to be construed to mean that such names may be regarded as unrestricted in respect of trademark and brand protection legislation and could thus be used by anyone.

Cover image: www.ingimage.com

Publisher:
Noor Publishing
is a trademark of
Dodo Books Indian Ocean Ltd., member of the OmniScriptum S.R.L Publishing group
str. A.Russo 15, of. 61, Chisinau-2068, Republic of Moldova Europe
Printed at: see last page
ISBN: 978-620-3-85998-0

Introduction to Renewable Energy Applications

By

Peyman Vaziri

Exploration Directorate of NIOC, Tehran, Tehran Province, Iran

Supervising Excavation Operations based on the Well Plan

Preparing Daily, Weekly, and Final Excavation Reports

Peyman Vaziri

Master of Science	**2017 – 2020**

- **Field: Oil and Gas Law**
- University of Tehran, Kish International Campus, Iran.
- Cum. GPA: 18.27 (out of 20)

Master of Science	**2010 – 2012**

- **Field: Petroleum Engineering**
- Petroleum University of Technology, Abadan, Iran
- Cum. GPA: 17.36 (out of 20)

Bachelor of Science	**2006 – 2010**

- **Field: Petroleum Engineering**
- Petroleum University of Technology, Abadan, Iran
- Cum. GPA: 15.47 (out of 20)

WORK EXPERIENCE

Drilling Supervisor   **2013 – Present**

Exploration Directorate of NIOC, Tehran, Tehran Province, Iran.

- Supervising excavation operations based on the well plan
- Preparing daily, weekly, and final excavation reports

Apprentice   **Summer 2009**

National Iranian South Oil Co. (NISOC), Shiraz, Fars Province, Iran.

- Acquaintance With Dehydration Units in Dalan Industrial Field
- Acquaintance With All Industrial Units of the Plant
- Working In the Drilling Unit and Acquaintance with Its Various Sections
- Acquaintance With Geology and Mudology Units

Apprentice   **Summer 2009**

Southern Zagros Oil & Gas Production Co. (SZOGPC), Shiraz, Fars Province, Iran.

Working in Services   **2005**

Fan va Bana Co., Jam, Bushehr Province, Iran.

Working in Services   **2005**

Pars Oil and Gas Company (POGC), Tehran, Tehran Province, Iran.

Activities, Honors, and Achievements:

2013 – Present

- Certificates in Drill string design, Wellhead design, Casing design, Technology of cement, Technology of drilling bits, well control-2, Drilling fluids and their applications, Rheology and Hydraulic of drilling fluid, Drilling fluid solid control, Fishing, Shales and their drilling problems, Artificial gas lifting, Full bore DST, HUET and Firefighting from Exploration Directorate of NIOC
- Certificate in IWCF level 2 from Aberdeen Drilling School
- Certificate in IWCF level 4 from Petroleum University of Technology

2007 – 2008

- Member Of Student Council of Abadan Faculty of Engineering
- Certificate in Petrel Software Package from TÜV, Germany
- Certificate in Geolog Software Package from Research Institute of Petroleum Industry (RIPI), Tehran, Iran

SKILLS

Operating Systems
- Windows XP/Vista/7 and 10

Scientific
- MATLAB-Pascal

Industrial Applications:
- Petrel, Vista, Geolog, Eclipse

Office Applications:
- Microsoft Office Collection, Internet Browsers

Chapter I

Introduction

Introduction

Currently, the world's electricity generation relies heavily on coal, oil and natural gas. Fossil fuels are non-renewable; they are based on limited resources that are gradually running out. In contrast, renewable energies such as wind and solar energy are constantly being replaced and never run out. Most renewable energies come from the sun, either directly or indirectly. Sunlight, or solar energy, can be used directly to heat and illuminate homes and other buildings, to generate electricity, to heat water, solar heaters, and a variety of economic and industrial applications. Pleasant heat also causes the wind to blow; The same energy received by wind turbines; Then the winds and the heat of the sun cause the water to evaporate. When this water vapor is turned into rain or snow and is directed from streams to rivers and waterways, its energy can be taken and its hydroelectric power can be used. Along with rain and snow, sunlight causes Let plants grow, the substance that makes those plants, we know as living mass or biomass. Biomass can be used to generate electricity, transport fuels or chemicals. The use of biomass for any of these purposes is called biomass energy. Hydrogen can also be found in many of the major constituents, such as water. Hydrogen is the most abundant element on Earth, but it does not exist as a natural gas. Hydrogen is always combined with other elements, such as its combination with oxygen to make water. When hydrogen is separated from its constituent element, it can be used as a fuel. Not all renewable energy sources come from the sun. Geothermal energy The geothermal valve for a variety of applications includes: generating electrical power and heating and cooling buildings, and the tidal energy of the oceans is due to the gravitational pull of the moon and sun on the earth. In fact, ocean energy comes from many sources. Also, the sun warms the ocean more than it does. It heats the surface, creating a temperature difference can be used as an energy source. All forms of ocean energy can be used to generate electricity.

Chapter II

Why is Renewable Energy Important?

Why is renewable energy important?

Renewable energy is important because of its benefits.

Its key benefits include:

Environmental Benefits: Renewable energy technologies are clean sources of energy that have far less contact and environmental pollution than conventional energy industries. Energy for our future generations: Renewable energy will never run out. But other sources of energy are limited and are dwindling these days.

Jobs and Economics: Investments in renewable energy are often spent on the supply of raw materials (supplies and goods) and consumption and structure for the construction and maintenance of equipment, rather than investing in costly imports of energy. This means that the money you pay for energy, instead of entering the economy of a foreign country, stays in our country, creates employment and saves fuel economy.

1- Environmental benefits

Renewable energy technologies are much more environmentally friendly than conventional energy industries that rely on fossil fuels. Fossil fuels play a significant role in many of the environmental problems we face today - greenhouse gases, air pollution and soil and water pollution - if the renewable energy component plays very little or no role. Do not have. Greenhouse gases, carbon dioxide, methane, nitrous oxide, hydrocarbons and chlorofluorocarbons surround the Earth's atmosphere like a warm, transparent blanket, allowing the sun's warm rays to enter and dissipating heat near the earth's surface. They trap (hold). The effects of this natural greenhouse keep the average ground temperature around 60 degrees Fahrenheit (33 degrees Celsius). However, increasing consumption of fossil fuels has significantly increased greenhouse gas emissions, especially carbon dioxide, which increases the effect of greenhouse gases, which are known as tangible and integrated heat of the earth. According to the US Environmental Protection Agency, the share of carbon dioxide is responsible for 1.2 to 2.3 times the general increase in temperature. However, renewable energy technologies emit little or no heat or electricity. Produces carbon dioxide. The use of fossil fuel energy is also an important source of air, water and soil pollution.

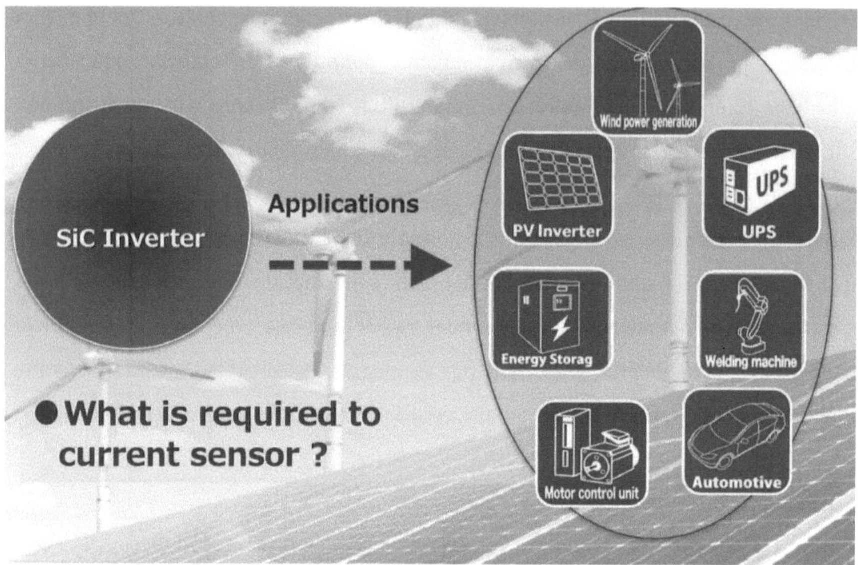

Figure 1. New ASIC Sensors optimized for SiC and renewable energy applications

Pollutants such as carbon monoxide, sulfur dioxide, nitrogen dioxide, particulate matter and lead - take a sad ransom from our environment! In other words, most renewable energy technologies produce little or no pollution. Both pollution and global warming pose a potential threat to the health of the human race. According to the Lung Association (USA), air pollution contributes to lung diseases such as shortness of breath, lung cancer, and respiratory infections, and about 335,000 people die each year in the United States. The long-term effects of global warming may also be more destructive. Complications of death are possible with very hot weather, and when the temperature rises, diseases can have a stronger latent energy for progress. Finally, renewable energy technologies can help us change traditional energy consumption patterns to improve the quality of our environment.

2- Energy for our future generations

What energy will the world consume in the future?

Yes, we can well prove that electricity consumption will grow globally. The International Energy Agency estimates that the world's electricity generation capacity will increase to approximately 5.8 million megawatts by 2020. Which is about 3.3 million megawatts, more than in 2000. At the same time, the planet's fossil fuel reserves, our current major source of energy, will begin to run out between 2020 and 2060, according to the best oil industry analysts.

How do we meet that need for that amount of energy?

Shell International predicts that by 2060, renewable energy will provide 60% of the world's energy.

The World Bank guarantees that the trade rate for solar energy (electricity) will reach a lump sum of 4$ trillion over 30 years. Biomass (biomass) fuels can also replace diesel. And unlike fossil fuels, renewable energy sources are sustainable and will never run out.

3- Jobs and economy

A large segment of the United States is forced to import fossil fuels, such as oil and natural gas, to generate electricity, heat, and fuel. These fossil fuels can cost millions of dollars, and every dollar spent on energy imports deducts a dollar from the local economy. At the same time, renewable energy sources are expanded locally, the cost of energy does not go out of the country, it creates jobs and strengthens the economy. Reducing renewable energy technologies takes a lot of effort. Jobs will soon end with the construction, design, installation, service and sale of renewable energy products. Employment will also be provided indirectly from jobs that supply renewable energy companies with raw materials, transportation, furniture and specialized services such as accounting and office services. As a result, the wages and salaries resulting from the job of the Haber increase the income in the local economy. After all, renewable energy revenues grow something more than this local economy, that is, benefits for the whole country.

In 2001, for example, the United States spent about $ 103 billion on oil imports. But as one of the world's largest manufacturers of renewable energy systems, it can bring more capital to its country by increasing the consumption of renewable energy around the world. At present, the manufacturers of US photovoltaic systems are about 2.3% of the world's manufacturers. And about 10% of the exports of these PV systems are mostly spent on development, which leads to annual sales of more than300$ million.

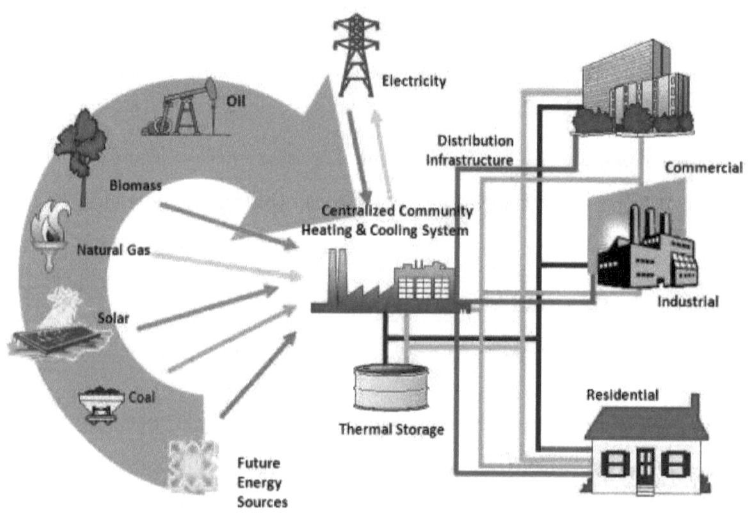

Figure 2. Hybrid renewable energy applications in zero-energy buildings

Why is energy optimization important?

Optimization means spending less energy on a single operation. Optimizing energy consumption in the country is by spending less money on energy by homeowners, schools, government offices, factories and industries. The money that should be spent on energy can instead be spent on consumer needs, education, services and products. An energy efficient economy can grow without consuming extra energy. An economy that consumes less energy and produces less pollution

Slow, because the two (energy consumption and pollution) are closely intertwined.

- ➢ **For homes:** For homes or small businesses and for other structures (efficiency) or energy optimization, less energy consumption means heating, cooling and lighting the building. It also buys energy-saving appliances such as computers and other home appliances. For homeowners and business owners, lower energy consumption is a financial reserve.

- ➢ **For cars:** For your car and other vehicles, energy optimization means building new trains and other vehicle technologies. Cars equipped with dual-fuel (gasoline-electric) engines or equipped with fuel cells are two examples of energy optimization in vehicles.

- ➢ **For power companies:** For power companies and other electricity (electricity) energy optimizers, the body often means helping their customers store energy in their homes and shops. Of course, it also means more efficient and better delivery and storage of electricity.

- ➢ **For local industries:** For local industries (limited and small industries), energy optimization means finding solutions that do the same job with less energy. Continuous casting, for example, in the steel industry, is an advancement in energy efficiency (optimization). Energy optimization also means better use of engines, steam systems, air compression systems and other industrial tools and equipment.

New energy

Here, renewable energies are categorized as:

- ➢ Living energy or biomass energy
- ➢ Live fuel
- ➢ Wind energy
- ➢ Solar energy
- ➢ Geothermal energy
- ➢ Hydroelectric energy
- ➢ Hydrogen energy
- ➢ Oceanic energy

Figure 3. SEI prepares students to empower developing countries with renewable energy applications

The place of solar energy in electricity supply

One of the questions we are justified in relation to energy is what will be the state of energy in the next few decades, what is the most economical source of energy and whether the sun can be considered as a source of energy with an economic profession. The need for energy is clear to everyone, and this need is becoming more apparent as technology advances and the world's population grows. Existing statistics show that the world's energy consumption is doubling in demand every 14 years. So far, the demand for electricity alone has almost doubled every 10 years, and this growth in demand in developing countries has accelerated, doubling almost every seven years. In the late 1980s, for example, US oil imports reached about 7 million barrels per day, which was almost double (1.3 million barrels per day) in 1980, while the amount of oil consumed to generate electricity in The United States accounts for about 4% of US consumption, and 52% of US electricity is generated from coal. Although coal is the major source of electricity in the United States, environmental regulations have been adopted to reduce acid rain due to coal burning pollution and global warming, and surface water pollution.

Sulfur causes an increasing population and technological advances that have improved living standards is a factor that will increase the demand for electricity. If we take a look at the demand for electricity from the world's existing resources from 1960 to 1990, we see that the use of different sources in the supply of electricity in the direction in which most of the fossil fuels have been used.

If we consider the rate of effect of electricity demand according to the table above 5%, the amount of energy requirement, which in 1987 was approximately 1.2×10^{14} kWh, in 2001, doubled to 2.4×10^{14}.

Obviously, considering these works, these sources cannot be relied on much and on the other hand, it will not be economical to use them, while in the same statistics, the amount of radiant energy absorbed by the sun is almost 100 times the resulting energy. It has estimated the existing fossil fuels. Therefore, more attention should be paid to solar energy in providing electricity. Now, considering these needs, energy sources and environmental problems caused by fossil fuels require more seriousness to find less dangerous energy sources. More desirable is being researched and studied Studies and studies have led to short-term and long-term strategies to provide electricity, in which solar energy plays a very important role because solar energy is practically unlimited and Pollution is available and the earth receives 7×10^{27} kWh of solar energy per year. These sources are: solar thermal energy, photovoltaic (neurolytic), geothermal (geothermal), biogas energy, etc. In these systems, neurolysis is one of the best methods of renewable energy that can be used in places with different climatic conditions. These systems can be used in rainforest deserts and in developed and developing countries. Neurolithic systems are systems that convert light directly into electricity. These systems are now mostly used independently of the national grid, but for applications that are far from electricity sources and due to the need and minimum maintenance of high capacity, no need for fuel and no contamination that can be expanded and installed. They are very efficient everywhere, and almost 97% of the neurolytic systems sold in 1990 are designed for off-grid applications. A Norlett system includes:

1- Solar modules

2- Battery

3- Charging the electroller

4- Consumers

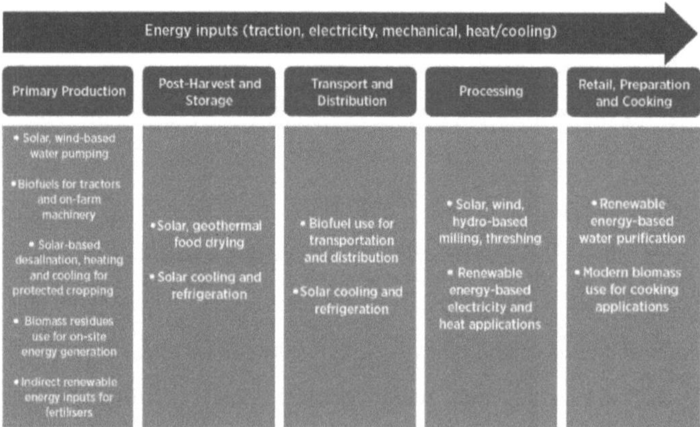

Figure 4. Renewable energy – European Biomass Industry Association

Solar modules

Modules or solar panels, which are the main power of a light system, are responsible for converting light into electricity. Solar panels are connected by a series of solar cells that are connected in parallel and in series.

Solar cells that are responsible for converting light into electricity.

Which are made of semiconductor materials and its different types are:

1- Solar cell made of single crystal silicon material

2- Solar cell made of polycrystalline silicon material

3- Solar cell made of amorphous silicon material

4- Solar cell made of silicone material

Here, the solar cell is made of polycrystalline silicon material.

A solar panel of type 36/45 MA is shown in which 36 solar cells are grouped together and have the following specifications.

Efficiency 11.5%

Current 7/2 amps

Dimensions 0.5×1 square meter

Voltage 16 volts

4.5-watt peak power under standard test conditions

Weight 5.5 kg

For applications with different voltages, the modules can be connected in series and in parallel.

Battery

Neurolytic systems produce electrical energy only if exposed to light, which is why batteries are used during night and cloudy days when the intensity of sunlight is low, and batteries when the intensity of radiation is appropriate by Solar panels are charged.

Charge controller

To protect the batteries, the voltage and amount of battery charge must be controlled, which is done by the electronic charging device of the controller.

DC to AC converter

The voltage generated by the solar panels is of the direct current type and if the consumer needs alternating current, a direct current converter should be used. As mentioned before, the high advantages of such systems and its high efficiency have led to the consumption of such systems to be considered.

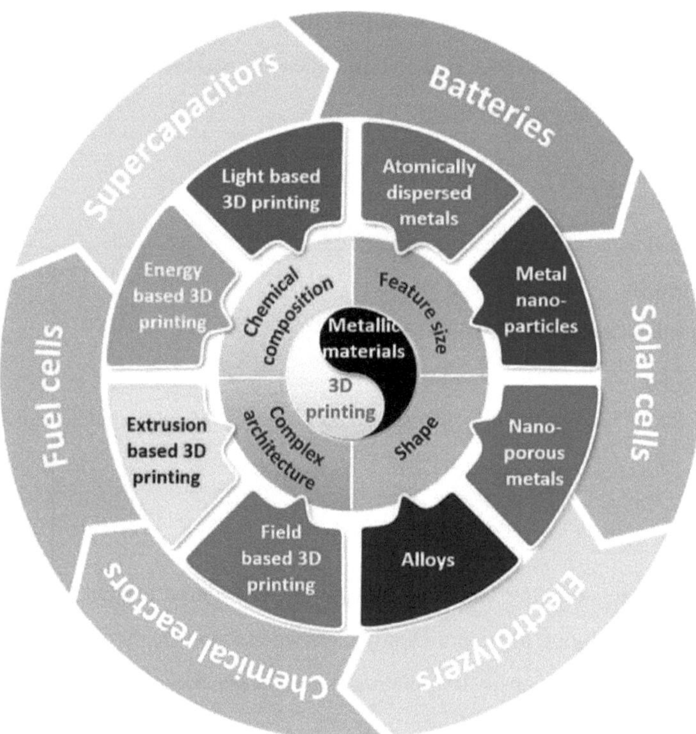

Figure 5. 3D printing of metal-based materials for renewable energy applications

Chapter III

Current and Future Position of Natural Energy

Classification of Solar Systems

Solar systems are the systems by which solar energy is used to meet the energy needs of human societies.

- o Photo biological systems
- o Chemical systems
- o Photovoltaic systems
- o Heating systems

Photo biological systems

The process of photosynthesis is the oldest and most widespread method of using solar energy. Plants absorb solar radiation and convert carbon dioxide and water into sugars. In the process of these interactions, plants release oxygen and absorb nitrogen and phosphorus, which are essential for their survival. The result of this process is the biological storage of solar energy. Solar energy (stored in plants) is recovered by burning wood or preparing fuels such as alcohol and methane. Today, the production of fuel from plant materials due to its low efficiency is rare.

The efficiency of this process is used between 0.25 to 0.5. It is a percentage that is significantly lower than the efficiency of other forms of sun use. Even with this low efficiency, the cost of producing energy from some plants is comparable to the cost of producing fossil fuels. It is a process of photosynthesis and is abundant in surplus agricultural and plant materials. It has been used as an energy source to provide food or chemicals needed by industry.

Solar chemical systems

Solar chemical systems fall into two general categories:

A) Photochemical systems in which solar radiation is used in chemical processes.

B) Heliothermic systems in which the sun is used as a heat source

Both systems are used in synthesis that require dual thermal and light energy, and as a result of this process, fuel is produced.

Photovoltaic systems

The process that converts solar energy into electrical energy without the use of moving mechanisms is called the photovoltaic phenomenon. The factor used in the process is called solar cells. The use of these cells about 43 years ago (since 1960) by using them as an electric generator in the solar field can produce solar energy with an efficiency equivalent to 5 Up to 20% can be converted directly into electricity, but their use has not yet been economical due to their high cost, except in difficult areas.

Solar cell function

Electricity is generated by properly constructed semiconductors. Today, the most effective and cheapest solar cells are silicon solar cells. Sand is one of the important sources of silicon and after its refining, silicon crystals are obtained. Which is layered after being cut. Pure silicon is a weak conductor in terms of electrical conductivity (semiconductor), but if a phosphorus is added to it during refining, a negative charge (electron) and if boron (Borne) is added, it becomes a positive charge (hole). The first type is called type N silicon and the second type is called type P silicon. Silicon has four electrons in the external circuit when we enter the number of phosphorus atoms into the system crystal, considering that phosphorus has five electrons in its external circuit. The four electrons in the outer circuit of phosphorus form one circuit with the four electrons in the outer circuit of the system, leaving one electron free and negatively charged silicon to form N-type silicon.

On the other hand, if instead of using the boron atom, which has three electrons in the outer orbit, spheres that can move like electrons are created, and silicon is positively charged. If we charge one side of a type P crystal of type N, a P-N bond is formed. On the P side there are free holes and boron atoms with negative and static charge, and on the N side there are free electrons and positively charged phosphorus atoms (due to the loss of one electron). There is a field created by static charged atoms. It is not possible because the holes on the P side cannot be replaced by the holes on the N side because there are no N holes. Similarly, the motion of the electrons on the N side is not. Now, if a photon (a particle of light) hits the transition surface, it can separate the electron from the silicon atom, creating a hole. Under the influence of the field, the hole moves to the P region and the electron to the N region, and these two opposing motions create different electric currents with different charges. Its return is a maximum of 20%. A photovoltaic cell is usually 300 microns thick and made of a circular silicon thickness of 3 to 9 centimeters in diameter.

The connections are on the outer surface of the cell, which is covered by a metal grid. The back of the cell is covered with a sheet of metal. There is another type of photovoltaic cell in which cadmium is used instead of silicon as the base element. The area of current cell surfaces has no effect on its voltage and is usually about 0.5 volts, but current intensity is a function of cell surface area and the intensity of solar radiation, and ideally 250 amps per square meter of cell surface to create a generator. The cells are connected to each other in series or in parallel. For example, if 24 solar cells are connected in series, a voltage equivalent to 12 volts is generated at a temperature of 25 °C. As the temperature increases, the productive power of the solar cell decreases.

SUSTAINABLE ENERGY REGULATION AND POLICY-MAKING FOR AFRICA

Wind Power Applications

Technology type	System	Application
Wind power - electrical	Grid connected	• Supplementing mains supply
Wind power - electrical	Stand-alone, battery charging	• Small home systems • Small commercial/community systems • Water pumping • Telecommunications • Navigation aids
Wind power - electrical	Stand-alone, autonomous diesel	• Commercial systems • Remote settlements • Mini-grid systems
Wind power - mechanical	Water pumping	• Drinking water supply • Irrigation pumping • Sea-salt production • Dewatering
Wind power - mechanical	Other	• Milling grain • Driving other, often agricultural, machines

Module 7

Figure 6. Renewable Energy Module

Uses

Photovoltaic cells are used in long-term designs due to their properties.

Photovoltaic cells do not need maintenance after installation. For this reason, it is used in places such as radio and television relay stations, naval traffic lights and air traffic lights, which are difficult to visit and take care of the system. In general, in places that are far from the energy grid and require a small but continuous amount of energy, the use of photovoltaic systems is economically and technically very appropriate.

Heating systems

It is currently the most economical solar energy system. This group of systems can be classified as follows:

A) Spa systems

B) Heating and cooling systems

C) Drying and cooking systems

D) Desalinate water systems

N) Pumping systems

F) Electricity generation systems

Flat solar collectors

The main element of solar bed collectors is a sheet that is heated by general sunlight and transfers its heat to a heat-absorbing fluid that is flowing. This fluid is usually water or air. The color of the sheet is always dark and it may have a special coating that maximizes the absorption coefficient of solar energy. Rubber, plastic and metal sheets are used for high temperature outputs. The system usually has a storage compartment to allow solar heat for night use.

If the system fluid is a liquid, the storage part is an insulator, and if the system fluid is air, some rock or concrete is used. This solution is a substitute, but in cases where they change phase, it is a better solution. But even with these advanced materials, it is still not practical to store heat for long periods of time, and as a result, most solar thermal systems use fossil fuel secondary systems as a complement to the system. In the collector circuit of this system, a solution of water and molecules is usually used. A water-to-water heat exchanger has been used to transfer heat from the storage tank to the building. An auxiliary heater is provided to provide energy to heat the space when it cannot be supplied by a tank.

Economic study of solar heating systems

Due to changes in weather conditions throughout the year, the use of systems that use solar heat will not be economical and it is better to use an auxiliary energy such as fossil fuels along with this system, this way The system will be able to operate in adverse weather conditions as well as in non-sunny conditions. In other words, the solar system will operate during the day (sunny hours) and fossil energy will be used in non-sunny times. In such a system, the cost of energy production or the ratio of energy and cost of consumption to the energy produced is less than a system whose total energy is provided by the sun. The costs required to build and maintain a solar heating system can be summarized as follows:

1- Cost of construction and installation of equipment

2- Fuel

3- Repairs and maintenance

4- System design

The equipment of the system itself includes different parts such as:

1- Collectors, supporting structures

2- Energy transmission system including pumps, pipes or ducts

3- Energy storage system

Investment

Decision making on investments is usually based on foresight and the amount of return that investment will bring in the future. Economic analysis should usually be based on the economic comparison of different alternatives, in which the cash flows of each alternative can be used as a criterion. Although engineering economics tries to imply quantitative parameters in the problem, it should not be overlooked that qualitative criteria overshadow quantitative importance in many cases. Social, cultural, environmental issues that are defined as project parameters. It may not be less important than the small amount of investment and return on projects. The usual type of investment in industrial units is a model of current investment and the cost or income that results from this investment. Usually, any investment is depreciated after what is called economic life and reaches a value called scrap value. In some cases, this value is considered zero. That is, the device or equipment has no value after its economic life.

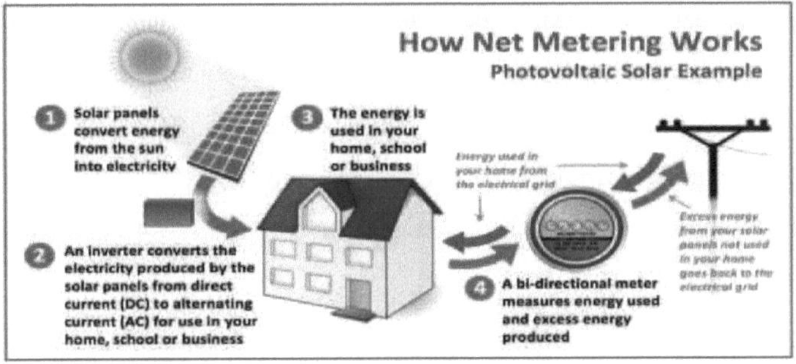

Figure 7. 4 factors driving renewable energy in Saudi Arabia

Initial cost

This amount is equal to the initial costs such as the cost of collectors, tank, pipe, installation cost and other costs that occur initially.

A) Collector cost

Collectors are divided into two categories. Water collectors have a higher cost than air collectors due to the presence of copper pipes for fluid transfer. The cost of water collectors will be 400,000 Rials per cubic meter, which is less than 200,000 Rials for air collectors due to the lack of copper pipes and personnel costs.

B) Cost of storage source

The storage source in most factories in Iran is made in different capacities, the cost of which is estimated at about 600 Rials per liter of source volume.

A) From the consumer's point of view

First, the use of natural gas as household fuel along with the solar system is examined. The price of each cubic meter of natural gas is 18 Rials. The average energy required to provide heating and hot water for homes (family of 5 people) in Iran is 121×10^3 mega joules per year. The average calorific value of natural gas is 37.3 mj / m^3 and the efficiency of gaseous fuels is considered to be 60%, so that the average annual consumption of each household will be 5500 cubic meters. If a percentage of the energy needed by each family is provided by solar energy. According to the efficiency of solar systems, the required area is obtained as above.

$$Q = I. A. \eta \rightarrow A = \frac{Q}{I. \eta}$$

In the above relation, Q is the amount of energy received, η efficiency and I the average annual average of solar energy on the sloping plate in the country, and for 34 points of the country, the average value of 6700Mj /m^2 .yr. The initial cost of solar systems, as mentioned, includes: collector cost, storage source, installation, structure and other costs, the cost of mass production for each square meter of collector is estimated at 300,000 Rials. For economic comparison of gas and hybrid fossil fuel systems (common gas and solar), the method of calculating the current value of costs has been used and the minimum absorption rates of 15 and 20% with a useful life of 25 years have been considered for the systems. For economic comparison of heating and spa systems from the household point of view and in general conditions, a relation can be provided based on the current value of solar and fossil system costs.

$$P_w = \frac{f.Q}{I.\eta}C + \frac{QR_f.(1-f)}{H_f\eta_f}\left[\frac{(1+i)^n - 1}{i(1+i)^n} + \frac{G}{100}\frac{(1+i)^n - 1 - ni}{i^2(1+i)^2}\right]$$

P_w = present value of system costs

f = fraction of the required thermal energy supplied by the sun.

Q = Total useful thermal energy per year

I = energy intensity per unit area per year

η_c = Collector efficiency

C = Initial investment cost (or price of solar system per cubic meter of collector)

H_f = calorific value of fuel consumed

η_f = Thermal efficiency of fossil fuel furnace

R_f = Domestic price per unit of fossil fuel

i = minimum capital uptake rate

η = Period or system life

G = Percentage increase in fuel prices in the first year

Therefore, taking into account the 20% increase in the base fuel price each year, the current value of costs for the fossil fuel system will be obtained.

Rials 1400000 = (Fossil) P_w

To provide 50% of the required energy by the sun and with an efficiency of 30% for the solar system, the required collector area will be obtained:

Rials 9700000 = P_w

$$A = \frac{0.5 \times 121 \times 10^3}{6700 \times 0.3} = 30m^2$$

The economic comparison of the two systems according to the current value of costs shows that the fossil fuel system has a lower value of current costs and is economical. The following results will also be obtained for a minimum absorption rate of 20%:

Rials 950000 = (hybrid system) P_w

Rials 954000 = (Fossil) P_w

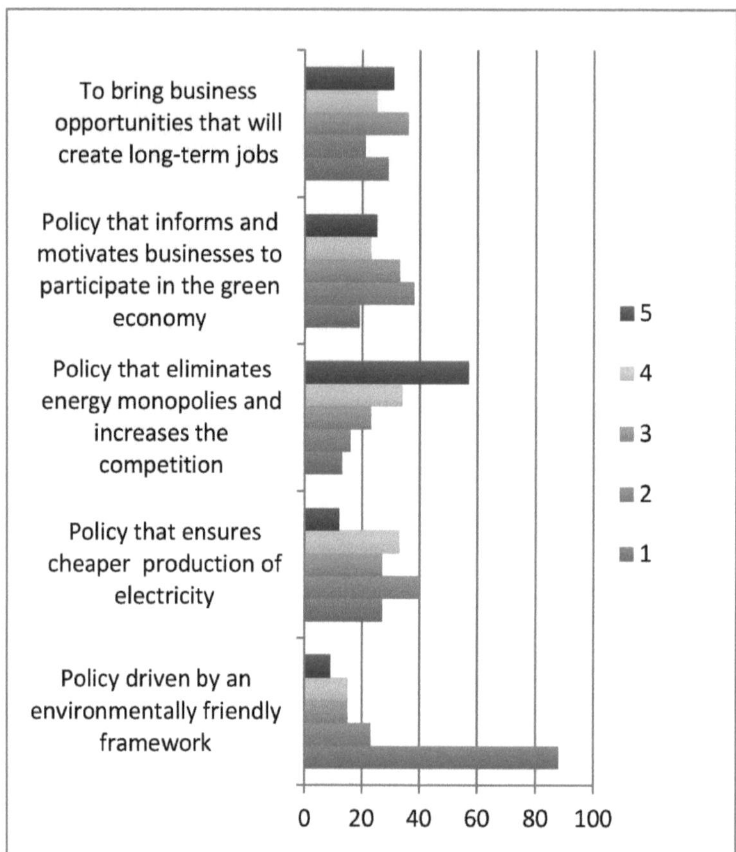

Figure 8. The Social Perspective on the Renewable Energy Autonomy

Again, an economic comparison shows that the solar hybrid system is not economical. The tables show that with increasing the efficiency of the solar system, the required collector level will be lower and the current value of the costs will be reduced. Compared to the fossil difference system, the current value of the costs will be lower and the economic conditions will be lower.

It is getting closer, so to increase the efficiency of solar collectors, higher efficiency technologies must be achieved with changes in the solar system. Table 4 also shows the decrease in the share of solar energy use, in which case the level of required collectors will be reduced and the current value of costs will be lower. Therefore, to reduce the difference between the current value of fossil and solar system costs, the share of solar energy use can also be reduced. This can be recommended at the beginning of the policy of expanding the use of solar heating and spa systems, and gradually increase the share of solar energy along with other policies. But this increase is also limited because there will be problems with the installation site because apartment living in Iran is expanding. In another study, diesel is used as a household fuel. The price of each liter of diesel, including the cost of transportation and 38 Mj/lite deliveries in homes, is 500 Rials. Will be equal to 6000 liters. Similar to gaseous fuels, the present value of costs will be calculated.

Rials 3,000,000 = (diesel fuel) P_w

And for a hybrid system with a share of 50% of solar energy we will have:

Rials 1500000 = (hybrid system) P_w

Economic comparison of the two systems according to the current value of costs shows that the diesel fuel system has a lower value of current costs and is economical. The following results will also be obtained for a minimum absorption rate of 20%:

Rials 10040000 = (hybrid system) Rials

2080000 = (diesel fuel) P_w

Again, an economic comparison shows that the hybrid system with diesel fuel is not economical from the household's point of view. In these figures, the dashed line shows the current value of costs for a 100% fossil fuel system. As can be seen, with increasing the efficiency of the solar system, the current value of costs decreases, but in all cases, the current value of the costs of hybrid systems is high and therefore the return systems have no capital and are not economical. Therefore, under these conditions and at the rates that the household pays for fossil fuels, these systems will be developed to the extent of research, demonstration of systems performance, improvement of systems and achieving high efficiencies and lower initial investment.

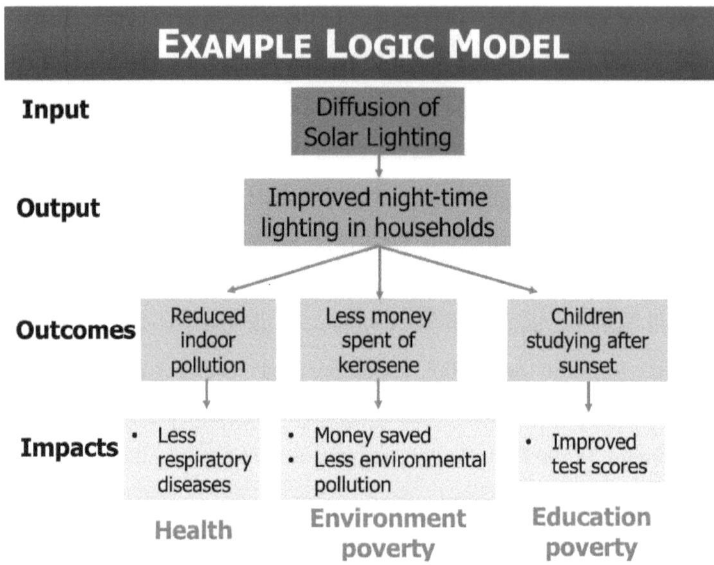

Figure 9. SEI prepares students to empower developing countries with renewable energy applications

B- Economic study from a national perspective

Internationally, international fossil fuel rates are used for economic research. For natural gas, the rate set for Iran's exports is 9 cents and for diesel, the import price is 14 cents, which has been studied for a value of 9000 Rials. According to a study by the World Energy Council, global fuel prices are projected to increase by about 3% per year, which has been used for economic comparison. Here, too, for the economic comparison of solar hybrid systems with fossil and from a national point of view in general terms, a relation can be provided based on the present value of the costs of the two systems. Pw = present value of system costs

$$p_w = \frac{f.QC}{I.\eta_c} + \frac{Q'R_f.(1-f)}{H_f\eta_f}\left[\frac{(1+E)^n(1+i)^n}{(1+i)^n(E-i)}\right] - \frac{QR_f.(1-f)}{H_f\eta_f}\left[\frac{(1+i)^n-1}{i(1+i)^n} + \frac{G}{100}\frac{(1+i)^n-1-ni}{i^2(1+i)^2}\right]P_w$$

f = fraction of the required thermal energy supplied by the sun.

Q = Total useful thermal energy per year

I = energy intensity per unit area per year

η_c =Collector efficiency

C = Initial investment cost (or price of solar system per cubic meter of collector)

H_f = calorific value of fuel consumed

R_f = Domestic price per unit of fossil fuel

E = rate of fuel increase per year

In this section, the current value diagrams of solar system costs in terms of efficiency are drawn. As can be seen, with increasing the efficiency of the solar system, the current value of costs decreases and in most current cases, the costs of hybrid systems are less than 100% fossil fuel systems, and therefore in these cases, hybrid systems return the return on investment They are fuel efficient and economical. These graphs can be used to evaluate the economics of solar systems in any situation.

Development policy of solar heating systems

In order to understand the effective contribution of the development of solar heating and spa systems in the future, we need to estimate the possibility of developing these systems and the cost of such development. The current state of these systems is a good starting point and we must look at vital parameters. Like current research objectives, it estimates the efficiency of materials, the improvement of materials, and the cost effects of mass production; however, not all of these studies are guaranteed because the possibility of predicting technological leaps and market structure of the economic situation in the future will be uncertain. To accelerate the development of performance demonstrations and the use of solar heating systems, the government can implement various types of incentive programs. These programs can include the following:

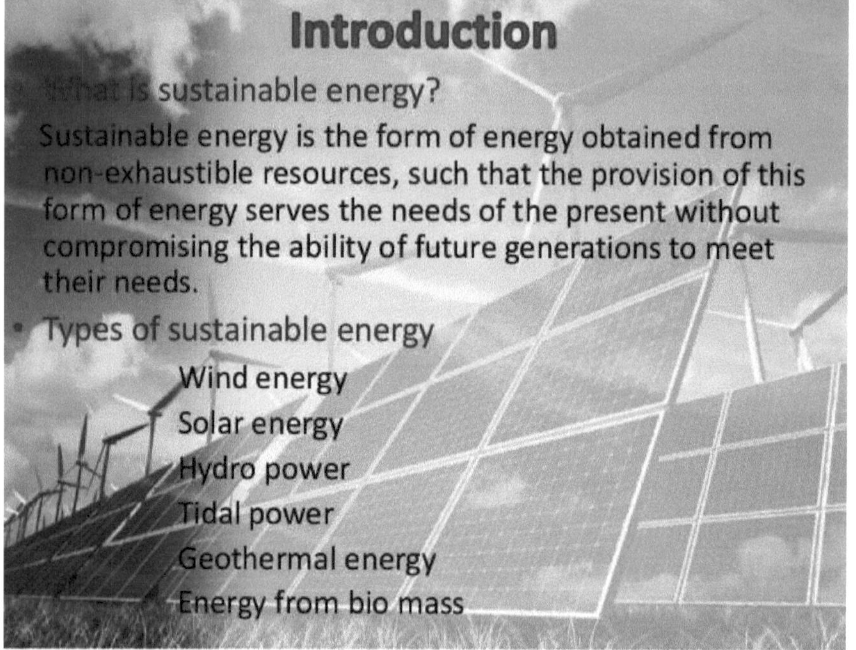

Figure 10. Sustainable energy applications in Agriculture

Economic assistance

This program eliminates some or all of the initial investment risks required for solar heating and hot water systems compared to fossil fuel systems. These grants can be in the form of investment tax exemptions, rapid depreciation schemes, low-interest loans, special concessions for building systems, subsidies for participation in investment costs for devices, or partial financing. The financial needs for the establishment of mass production facilities should be met by the government.

Research, development and demonstration of systems performance

In these programs, government investment can be used to contribute to the costs of demonstrating the performance of systems before there is a demand in the market, and in this case, the manufacturers, the minimum financial risk of program delays and development costs in a market Rapidly expanding, they will be willing to participate.

Mandatory use of solar heating and demand control systems for fossil fuels

Mandatory bylaws for new commercial and residential buildings or government facilities indicating the use of solar heating and hot water systems, the demand for solar heating systems can be met by preventing people from using fossil fuels through taxes or Quota setting increased. Other incentive programs include paying long-term, low-interest, tax-free special loans by banks to invest in solar hot water systems. In an income-based economy, investment tax cuts and rapid depreciation assistance are effective tools to accelerate the use of solar heating systems because:

A) Reduces the effective cost of investment to amounts that are comparable to fossil fuels.

B) Private companies raise funds to finance performance demonstration projects and build commercial construction capabilities. Using the experiences of other countries in the use of solar systems will be very effective. These experiences and information are available but should be widely disseminated, and future experiences require a good summary of documents and information and separate independent activities. Incentive programs can and should be developed to achieve very good and wide final applications, and efforts should be made to overcome some of the important factors that tend to limit the use of solar heating systems.

Some of these restrictions include:

- Economic competition with fossil fuels
- Initial investment cost and future operating cost
- Total capital required for investment
- Reasonable growth in mass production
- Needed changes in lifestyle and education

To eliminate the effects of these limiting factors, some programs that have the following characteristics should be implemented by the government wherever necessary:

Technical

- ➤ Develop local and regional databases to assist with planning
- ➤ Invest in research and development to acquire technology talent and effectively engage and expand RSD into the private sector
- ➤ Applying the results of research and development programs in different cities
- ➤ Emphasis on the design and construction of appropriate quality systems

Economic

- ➤ Directing financial resources to the private sector to build these systems
- ➤ Examining the external costs of fossil fuels when setting and determining subsidies
- ➤ If possible, incentives for energy delivered by these systems in addition to investment costs.

Education / Social - Cultural

- ➤ Linking programs to savings initiatives
- ➤ Education and information for the consumer / public
- ➤ Offer free evaluation services to consumers by trained professionals.

Multiple uses of solar energy

As mentioned, solar energy is the total energy on the planet. To understand the subject and the energy of the sun from the perspective of astronomical sciences, and how to use each of them is given below.

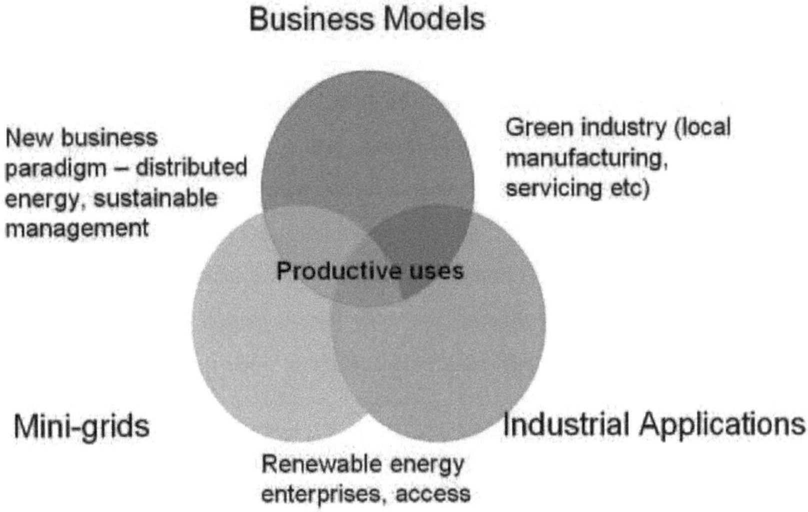

Figure 11. Renewable energy | UNIDO

1- Astronomical Sciences

From the point of view of astronomical sciences, solar energy is divided into two forms: radiant energy and thermal energy.

A) Use of radiant energy

It is mostly used for electricity generation and lighting. The amount of radiant energy of the sun on the earth is 10,000 times the annual amount of energy consumed by the whole world. Photovoltaics can be used to convert radiant energy into electricity took advantage of it and converted it directly into electrical energy. This method can be used to supply electricity to a village. Its technology is simple and maintenance is not necessary. For example, photovoltaics generates electricity at 100 W / m^2 with 15% efficiency in sunny weather. The illumination of the sun's radiant energy environment takes place naturally and the phenomenon of illuminating the earth during the day is caused by radiant energy.

B) Thermal energy

The solar thermal energy itself has various direct uses. For example, using solar collectors can be used for spa, indoor heating, cooling, fresh water, drying of agricultural products and electricity and heating.

For example, in comparison with cooling and heating with conventional systems, using solar collectors and absorption chillers, it is possible to save 52% and 39% of energy required for heating and cooling houses, respectively, or by increasing the level of the lens and the more radiation collected, the more heat it gained. In France, the solar furnace mounted on the Pyreness has a power of about 3300 degrees Celsius, which can melt tungsten metal.

Environmental sciences

Solar energy is divided into two types from the point of view of environmental sciences: water energy and wind energy.

A) Rivers that are created by the heat of the sun and the melting of snow. They can be effective in generating electricity in two ways. First, electricity generation by water energy is possible by dropping water from a great height and moving the turbine. In this method, construction of dams is a clear example of this example. Secondly, the movement of river water also allows the turbine to rotate, which in turn generates electricity.

B) The wind that arises from the heating of the air layers and the movement of this air and its displacement. It is now used to generate electricity and heating, and the United States and Western Europe in the early 1980s tested the production of several megawatts of electricity from wind power. Of course, so far in the production of electricity by wind power, by installing a set of wind turbines in one place has succeeded in obtaining 50 to 100 kW of electricity, especially in the states of California on the West Coast by installing 10,000 wind turbines with a total capacity of 1000 kW Electricity is currently being generated.

Chemical sciences

From the point of view of chemical sciences, we can name chemical interactions with light. In photosynthesis, plants convert light energy into chemical energy. Chlorophyll in plant cells absorbs sunlight and, by increasing the energy in them, causes chemical interactions, and finally carbon dioxide (co_2) and water (H_2o) and Oxygen (o_2) is reduced. This energy not only promotes the growth and maintenance of plants but also plays a major role in human and animal life on Earth. These interactions convert carbon dioxide (co_2) into oxygen (o_2) and keep atmospheric compounds uniform and balanced. According to some experts, oil, coal, gas, and in other words fossil energy over time have been created by photosynthetic energy stored in plants (this theory has not been accepted by some other experts).

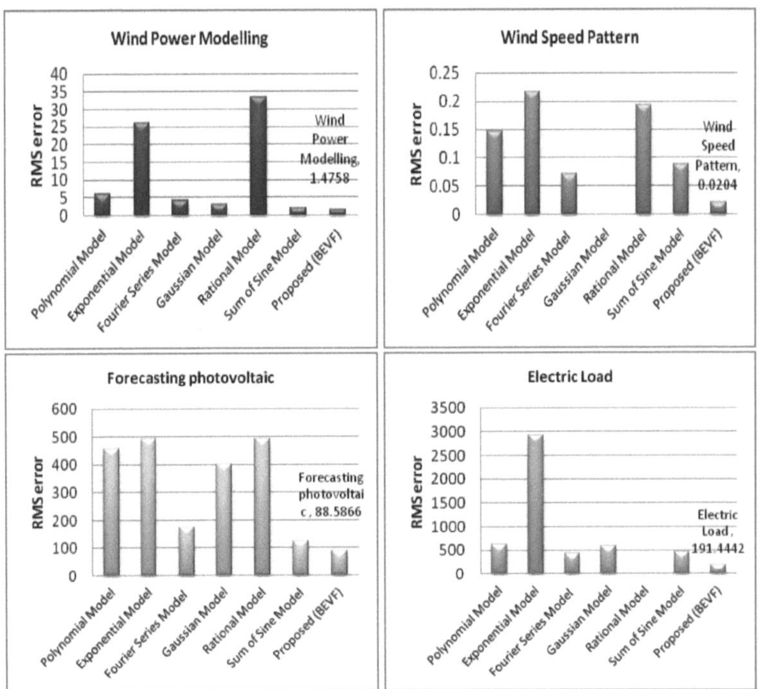

Figure 12. PLOS ONE: Field data-based mathematical modeling by Bode equations

Chapter IV

Solar Constant

Solar Model

The sun is the ultimate source of most of the energy currently available to the earth. This energy includes energy for direct heating, wind energy, hydroelectric power and energy from fossil fuels. Existing fossil fuels are the result of photosynthesis. The process by which plants convert solar energy into chemical energy. A full understanding of solar energy technology is possible only through a thorough analysis of solar radiation.

The sun, the star closest to us, generates energy to survive on Earth, and creates the gravitational pull required to keep our planet in an almost circular orbit. The sun has a mass of $M=1.99\times{}^{30}10$ Kg (approximately 3.3×10^5 times the mass of the earth) and radius is $R = 6.96\times10^8$ m (approximately equal to 109 times the radius of the earth). The distance between the earth and the sun is from 1,0167 astronomical units (at an unlikely solar point, approximately 13 July) to 0.983 astronomical units (one astronomical unit is approximately 1.5×10^{11} meters). The inner part of the sun is not available to us for direct experiments on it, but based on observations from the surface of the sun and theoretical studies, astronomers believe that its inner temperature is about 15 million degrees Kelvin, the chemical composition of the sun in Mostly hydrogen and less helium. These two chemical elements, which make up 96 to 99 percent of the sun's mass, are under intense pressure, and only the sun's high gravitational pull holds the mass together. Energy inside the sun is produced by nuclear fusion of hydrogen to helium.

This energy makes its way to the surface of the sun and eventually emits mainly in the form of electromagnetic radiation in space. The surface of the sun or photosphere is actually the transition zone where density decreases rapidly. By passing the sun to the outer part of the photosphere, we reach a relatively transparent environment in terms of light. In addition, the temperature drops to about 6,000 degrees Kelvin. At the top of the photosphere is the atmosphere of the sun, which is called the chromosphere because it absorbs certain colors of radiation from the photosphere because it is relatively transparent, its effect on solar radiation is ignored. Most of the radiation that reaches us is emitted from the photosphere, and therefore the solar spectrum is determined by the optical and thermal properties of the sun's surface. In the simple model used here, it is assumed that the sun behaves like a black body with a surface of T≅6000 $\overset{\circ}{o}$ ^{k} constant. This surface temperature is kept constant by an energy source inside the sun. Due to this high temperature, the surface of the sun emits light and emits electromagnetic radiation in all directions of space.

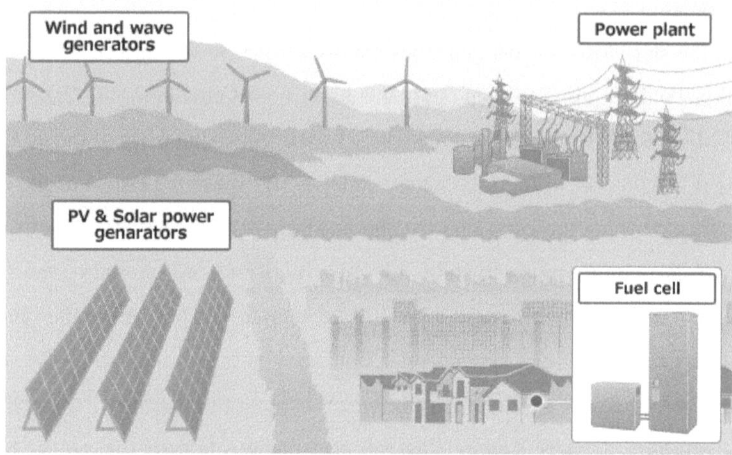

Figure 13. Power Plant and Renewable Energy

Blackbody Radiation: Electromagnetic radiation is made up of waves of oscillating electric and magnetic fields. Each wave is characterized by wavelength and frequency v. In a vacuum, all waves travel at the same velocity equal to $C = 2.9979 \times 10^8$ meters per second. Frequency, wavelength and λ velocity v of each wave are related according to the opposite relation: $\lambda v = C$

Only a very narrow band of wavelengths in the range of 900 nm< λ <400 nm is visible to the human eye. The wavelengths at the boundary of the visible spectrum at the end of the violet region (λ <400 nm) are called the ultraviolet wavelengths and are invisible at the wavelengths at the red border (λ > 700 nm) They are also invisible. As we will see later, about half of the solar radiation is in the infrared, while the composite components of the visible spectrum make up less than 40% of the sun's energy. When electromagnetic radiation shines on the surface of an object, it can be transmitted, reflected or absorbed. If the body is cloudy, no transfer is possible. The radiant energy at a single time at a single wavelength that is irradiated on the surface is called the spectral flux $F_\lambda^{(i)}$. Similarly, the absorbed and reflected spectral fluxes are denoted by $F_\lambda^{(a)}$ and $F_\lambda^{(r)}$, respectively. The index λ indicates that we are dealing with a single component wavelength. The total flux in the distribution diagram is as follows:

$$F = \int_\infty^\infty F_\lambda d\lambda$$

Spectral absorption, α_λ and spectral reflectivity r_λ, the surface of an object are defined by the following relations:

$$a_\lambda = \frac{F_\lambda^{(a)}}{F_\lambda^{(i)}}$$

And

$$r_\lambda = \frac{F_\lambda^{(r)}}{F_\lambda^{(i)}}$$

When the body is cloudy, what is not reflected from its surface must be absorbed and we can write:

$$a_\lambda + r_\lambda = 1$$

In fact, $a_\lambda \ and \ r_\lambda$ for a real surface, they depend on the wavelength of the flux and the direction of the radiation. For example, many surfaces absorb reflected light well in the vertical mode, whereas they are not effectively absorbed when the light is projected at near-vertical angles. We ignore this dependence on the direction of the radiation and, for convenience, assume that the surface is a matched absorber. Nevertheless, the reflectivity and absorption spectrum vary considerably along the wavelength of the radiation flux. Many pigments turn white due to good reflection in the visible spectrum.

Many absorbers are exposed to infrared radiation. The following definitions are useful based on the idealization of real surfaces:

Blackbody: Any object whose surface absorbs all the electromagnetic radiation components, regardless of wavelength or direction of radiation, is called a blackbody. In such objects: for all wavelengths

$$0 = r_\lambda \qquad\qquad 1 = a_\lambda$$

White matter: Any object whose surface reflects all the components of electromagnetic radiation regardless of wavelength and direction of radiation is called a white body (or full reflector). In these objects:

For all wavelengths $\qquad\qquad (1 = r_\lambda)\ 0 = a_\lambda$

Gray body: Anybody whose surface absorption capacity is between the surface absorption of a black body and the surface absorption of a white body and is independent of the wavelength and direction of radiation is called the gray body (matched). In such objects: that $0 < \alpha < 1$ (for all wavelengths)

$$a_\lambda = a$$

In fact, no surface is completely white or black. Black polish against solar radiation has an average absorption capacity of a = 0.97 and polished silver which is very shiny has a = 0.07.

It is an experimental fact in nature that whenever a turbid body is kept at a constant temperature, its surface emits a special electromagnetic radiation called thermal radiation. This radiation is generally propagated in all directions and includes all wavelengths of the electromagnetic spectrum. The heat flux that leaves the body depends on the surface properties of the body as well as its temperature (T in Kelvin). For matched adsorbent surfaces, the radiated heat flux is matched and its spectral distribution is as follows:

$$F_\lambda = \varepsilon_\lambda B_\lambda(T)$$

In this:

The formula ε_λ is a surface property called the spectral emission prior and $B_\lambda(T)$ is called the Planck function. This general function λ and T are obtained from the following formula:

$$B_\lambda(T) = \frac{a}{\lambda^3(e^{\lambda ht} - 1)}$$

Fixed values in this function are as follows:

$$= 1.4388 \times 10^{-2} \text{m-k}$$

$$a = 2\pi hc^2 = 3.7405 \times 10^{16} \text{ W-M}^2$$

where in:

(Joules per second) $6/252 \times 1^{-34}$ j-s = (Planck constant) h

$$2/9979 \times 10^8 \text{ m / sec} = \text{(speed of light) c}$$

$$1/3806 \times 10^{-23} \text{ j / k} = \text{(Boltzmann constant) k}$$

(Jules Berkeley)

The total radiation flux emitted by the surface is:

$$F = \int_o^\infty \varepsilon_\lambda B_\lambda(T)d\lambda$$

According to the laws of thermodynamic equilibrium, it can be proved that the ability to emit heat and the ability to absorb light are actually related. This relationship is established according to Kirchhoff's laws. The spectral emission capacity of an identical surface is equivalent to its spectral absorption capacity, or in other words:

$$\varepsilon_\lambda = a_\lambda$$

From Equation conclude that the black body in which it is $a_\lambda = 1$ located $\varepsilon_\lambda = 1$ is the most efficient emitter for all waves. Hence, for a black body, Equation looks like this:

$$F_\lambda\ \text{Black} == B_\lambda(T)$$

So that the function that describes the spectral flux emitted from a black surface at Kelvin temperature is the same as the Planck function. From Equation 3.5 we also find that a white body does not emit any heat radiation and a gray body emits radiation according to the following equation:

$$F_\lambda\ \text{Gray} = \quad (0 < \varepsilon < 1) \quad = \varepsilon B_\lambda(T)$$

A real object is drawn at 6000 degrees Kelvin. Note that the spectral function for a black body is equivalent to the Planck function, while the spectral function is the same for a gray body and ε is only small in size. Therefore, it is important to pay attention to $B_\lambda(T)$ mathematical properties. The area below each curve is finite and each has a wavelength at which the value is maximal. Therefore, at any finite temperature, the energy carried by components whose wavelengths are too short or too long is so small that it can be ignored. In addition, most of the energy is displaced by wavelengths in that region. Which has $B_\lambda(T)$ its largest value.

It can be shown that the following mathematical properties $B_\lambda(T)$ are valid:

$$(\text{Law of Movement})\ \lambda \frac{a}{T_{max}}$$

$$(\text{Stephen-Bonterman Law})\ F = \int_{o}^{\infty} B_\lambda(T)\,d\lambda\,\delta T^4$$

Where a and δ are universal constants.

According to Equation, it is known that the wavelength at which it has its $B_\lambda(T)$ maximum value, inversely, changes the Kelvin temperature. According to Equation, we see that the total area below $B_\lambda(T)$ and consequently the total heat flux emitted by a black body is proportional to the fourth power of Kelvin temperature. In the following, we express all the quantities in the MKS (meter-kg-second). In this device, energy is expressed in joules (j), power (energy in time) is expressed in watts (w) and area is expressed in square meters. In addition to these units, the total flux is expressed in watts per square meter and the wavelength is easily expressed in meters, microns $(1\mu m = 10^{-6}m)$ or nanometers $(1nm = 10^{-9}m)$. The value is usually said to λ_{max} represent the "color" property for a given planck distribution, although it is not necessarily the color that is recognized by the human eye. But the fact that λ_{max} it decreases with increasing T explains why a black body becomes so hot at a certain temperature that it turns red and turns white as the temperature increases further. White eyes indicate the presence of blue components. Using Equation, we find that a black body at 7000 K has a wavelength with a characteristic (blue) of $\lambda_{max} = 0.414\lambda m$.

While another black body at T=5800° ^{k} has a wavelength (green) $\lambda_{max} = 0.5\lambda m$. If a black body is at room temperature (T=300° ^{k}), then (infrared) will be $\lambda_{max} = 966$ λm and the black body will be visible. The total flux emitted from a black body according to the following formula is obtained from Stephen-Buttermann's equation:

$$\text{black } F = T\delta^4 = (5.67\times10^{-8})T^4$$

Hence, black bodies with surface temperatures of 7000, 5800, and 300 K, respectively, emit fluxes of $1.36\times10^8, 6.4\times, 10$ and 4.59×10^2 d watts per square meter, respectively. Note that as the Kelvin temperature rises, there is a significant increase in blackbody emissions. Because the emission of a black body changes in proportion to T^4, doubling the value of T increases the value of black F by 16 times.

Emission of radiation from the sun

Here if the solar model is assumed to be a black body at constant temperature T. In this case, the flux emitted from the surface of the sun can be represented by a Planck distribution. The spectral distribution of the observation of the sun $B_\lambda(T)$ is slightly different because the sun is neither in a state of radiative equilibrium nor even in a permanent state.

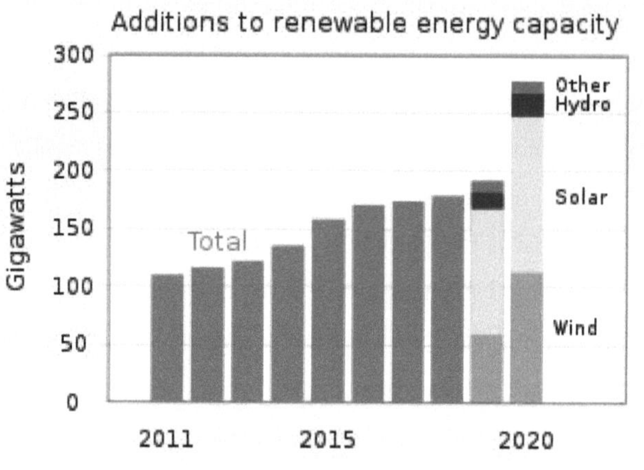

Figure 14. Renewable energy

Nevertheless, the curve of an object corresponding to a temperature of T=5800˚ k is my approximation for the solar spectrum. From now on we will use this black body approximation. Using Equation, we see that the characteristic wavelength of the solar spectrum is:

$$\lambda \frac{2.9 \times 10^3 \mu m - k}{5800k} {}_{max}$$

Which is equivalent to green light. From Equation, we find that the total flux coming out of the surface of the sun is:

$$F_o = \delta T_o^4 = \left(\frac{5.67 \times 10^{-8} w}{m^2 - k^2}\right)(5800k)^2 = 6.416 \times 10^7 w/m^2$$

This radiation is scattered after rising from the surface of the sun (it is scattered in all directions). The total radiative power emitted from the sun is obtained by multiplying the above flux by the area of the sun's surface, we have:

$$P_\circ = F_\circ 4\pi R_\circ^2 \cong (6.42 \times 10^2 w/m^2)4\pi(6.69 \times 10^8 m)^2 \cong 3.91 \times 10^{26} w$$

If the sun emits its radiation in the same way, this huge force, which astronomers call radiance, will be emitted equally in space in all directions. As the distance from the sun increases, this power spreads over the spherical surfaces, which have an increasing area, so that the light intensity is inversely proportional to the square of the distance from the center of the sun. At distance r is the area of the surface $4\pi r^2$, so that the flux of radiation that reaches such a surface is:

$$F = \frac{P_\circ}{4\pi r^2} = \frac{4\pi R_\circ^2 F_\circ}{4\pi r^2}$$

$$F = R_\circ^2 F_\circ / r^2 \cong \frac{3.11 \times 10^{25}}{r^2} w/m^2$$

Because the Earth's distance from the Sun changes throughout the year, the flux reaching the Earth changes doing. In the average distance from the earth to the sun, i.e. r=1.5×10^{11}.

The flux value is:

$$S = F \cong \frac{3.11 \times 10^{25}}{(1.5 \times 10^{11})^2} = 1382 w/m^2$$

This amount of flux, known as the solar constant, and previously explained, is not really a constant and varies with the seasons and to some extent with solar activity. In addition, note that the numerical value of Equation, is obtained by assuming that the solar spectrum, as a black body, is approximately equal to 5800 Kelvin. If we wanted to change this temperature to 5762 K, the exact calculation would show that the solar constant would drop to approximately 1352 watts per square meter.

The constant amount of solar has been measured by different researchers and varies between 1970 to 2013 watts per square meter, the difference in results is about two percent. Assuming that the temperature of the spectrum is approximately 5760 K, we optionally choose 1352 watts per square meter as the constant solar value. Equation, is also valid for spectral distribution and we can write:

$$S_\lambda = \frac{R_\circ^2}{r^2} F_{\circ\lambda} = \frac{R_\circ^2}{r^2} B_\lambda(5760^k)$$

$$\cong 2.156 \times 10^{-5} B_\lambda(5760k)(w/m\text{-}m)$$

With:

$$s = \int_\circ^\infty S_\lambda d\lambda = 1352 w/m^2 - m$$

As a result, the spectral distribution of flux when it reaches the Earth's atmosphere is necessarily the same as the spectral distribution emitted by the sun. However, each component of the spectrum attenuates equally as it passes. However, the emitted flux that leaves the surface of the sun is scattered. But when it reaches the atmosphere, it reaches almost unidirectional or beam-like.

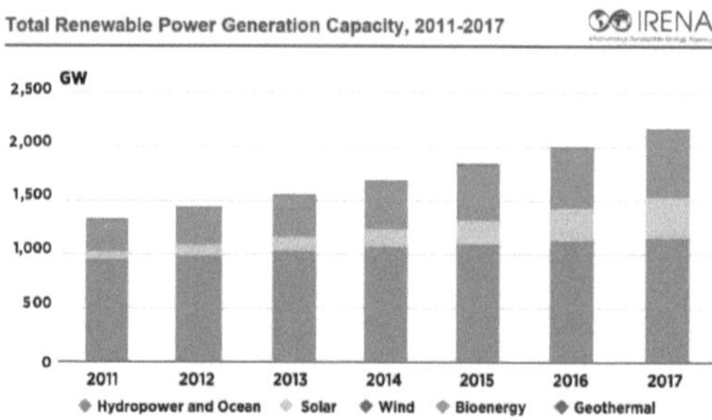

Figure 15. Global Renewable Generation Continues its Strong Growth, New IRENA Capacity Data Shows

As it moves away from the sun, its disk appears to shrink and all its energy comes from a very specific direction. Because the sun is not exactly a point but looks like a disk, its radiation is not entirely radiant and is somewhat divergent.

Therefore, the angular divergence is:

0.53= Varian

$$\Delta\theta = \frac{2r_\circ}{r} = \frac{(2)(6.96 \times 10^8 m)}{(1.5 \times 10^{11} m)} = 9.28 \times 10^{-3}$$

This value $\frac{1^\circ}{2}$ is slightly higher than so that the flux forming a constant solar can be considered as one-way radiation.

Solar fixed spectral composition

As mentioned, the fixed solar spectrum is determined by Equation 1-12, which is as follows:

$\times B_\lambda(T)$ fix $= S_\lambda$

The fraction of energy (F) transmitted by wavelengths between 0 and λ is proportional to the area under the black body curve between the two limits. This value can be determined by solving the following integrals:

$$f_\lambda(T) = \frac{\int_\circ^\lambda B_\lambda(T)d\lambda}{\int_\circ^\infty B_\lambda(T)d\lambda} = \frac{\int_\circ^\lambda ad\lambda/[\lambda^5(e^{b\lambda l}-1)]}{\int_\circ^\infty ad\lambda/[\lambda^5(e^{b/\lambda T}-1)]}$$

Equation shows that the integrals can be solved and evaluated for any temperature without exception. This is not something that can be seen by changing the integral $X = \lambda T$ using substitution. Using Equation, the relevant equation can be written as follows:

$$f_\lambda(T) = f(\lambda T) = f(X) = \int \frac{adx}{\delta x^5(e^{b/x}-1)}$$

The amount of this fraction for:

$$x_1 = \lambda_1 T = (0.4\mu m)(5760k) = 2304\mu m - k$$

Note

This fraction will be for the same wavelengths at a lower temperature of 3000 degrees Kelvin.

$$[x_1 = (0.4)(3000) = 1200\mu m - k] \quad \text{Only } f(1200)_{\mu m - k} = 0.002 = 0.2$$

For example, to find the energy fraction for wavelengths between micrometers

$$\lambda_1 = 0.4 \text{and} \lambda_2 = 0.7 \text{ for T} = 5760^{\text{ k}},$$

we use the following expression:

$$F(4032m\mu\text{-k})\text{-f}(2304m\mu\text{-k}) = 0.488\text{-}0.121\text{=}0.367\text{=}37\%$$

The residual fraction, which is determined by wavelengths higher than $= 0.7 \ m\mu$, is obtained using the following equation:

$$1\text{-}f_{\lambda_2}\text{=}1\text{-}0.488 \cong 0.51 = 51\%$$

If we approximate the solar spectrum by distributing a black body at 5760 K, then, as we have already seen, approximately 12% of the energy is transmitted by wavelengths shorter than 0.4 $m\mu$. This action is mainly in the form of ultraviolet radiation. The visible part of the solar spectrum contains 37% of the energy, while the wavelengths longer than 0.7 $m\mu$ (mostly infrared) contain 51%.

Therefore, approximately two-thirds of the energy received from the sun is invisible to the human eye. In short, the solar flux that reaches the upper layers of the Earth's atmosphere is primarily electromagnetic. Its spectral distribution is quite similar to the spectrum emitted by a black surface at 5760 Kelvin. $\frac{1}{2}$ This energy almost reaches the earth in the form of infrared rays and $\frac{1}{3}$ it is located in the visible region of the spectrum.

This flux is basically radiant or in the form of unidirectional radiation whose angle of divergence $\frac{1}{2}$ is approximately degrees. The total flux (average of the seasons) that shines on a surface facing the sun is called the solar constant and is numerically approximately equal to:

$$S = 1352 w/m^2 = \frac{1.94 Ly}{min \frac{429 Btu}{hr - ft^2}}$$

Which: 1=252 cal=2/929×10⁻⁴Kwhr British unit of heat (Btu)

And: 1cal/cm² = 1 length (Ly)

To make predictions about the abundance and availability of solar energy on Earth, we need to consider the apparent motion of the sun on the celestial sphere. The subject of the sun during the day, as well as the length of the day itself, both determine the amount of solar energy that is provided to solar collectors.

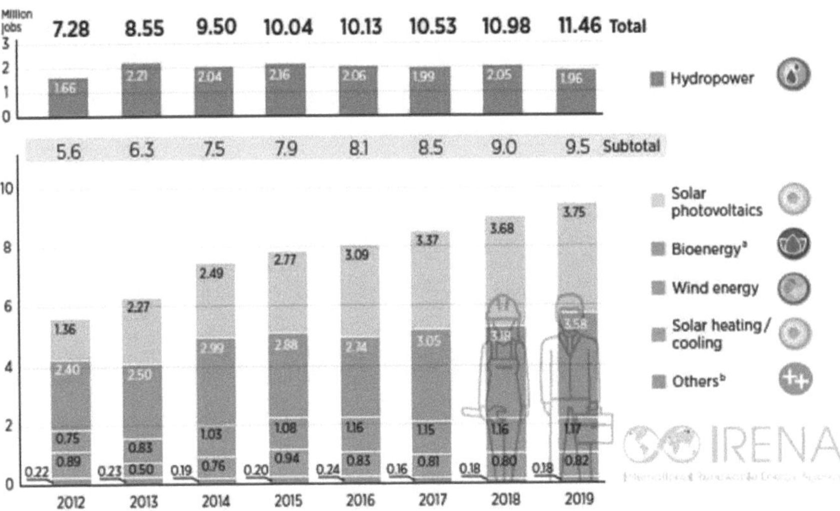

Figure 16. Renewable energy jobs continue growth 2019 to 11.5 million worldwide

Chapter V

Solar Heating Systems

In This Chapter, We Consider the Solar Heating System in General.

In particular, we will look at how to put a string of solar panels together and connect them to other devices to have a complete, efficient and cost-effective heating system.

Orientation of panel strings

Flat panels, unlike concentrators, which usually require a daily tracking system, are able to work with a fixed orientation. Tracking, however, will improve the performance of a flat panel. But the profit usually corresponds to the higher costs incurred in the construction and maintenance of the tracking device.

A fixed row of flat panels should be placed in a direction that has the maximum daily received flux during the working season. Since it is difficult to determine the best angle of inclination relative to the sun's diffuse radiation (if not impossible), we will obtain the best value for the direct component. Using the following equation, we obtain the daily direct flux, which is obtained by a string with skew coordinates Δ and ϕ:

$$(1\text{-}4)\ (\Delta, \phi) = S \int_{1}^{2-\tau/csz} ec\theta dt$$

In this formula, S is equal to 1352 w/m^2 (solar constant) and τ is the optical thickness of the Earth's atmosphere.

Z is the angle of the solar head, θ the angle of inclination of the sun's rays towards the set.

t_1 and t_2 are solar times between which $C\theta$ is positive. That is, when the sun's rays shine on the front surface of the string. ϕ And Δ the angle of the side and the angle of inclination are two strings, respectively. These values are for a set in the complement of latitude L= 49˚ and $\tau = 0.3$

the optical thickness of the atmosphere = 0.3. It is chosen that at sunset the set is approximately perpendicular to the sun's rays, i.e. when:

$$= |D' - L'|_{\text{Noon}}\ \Delta \cong Z$$

During the winter solstice, a vertical string facing south is more effective than a horizontal string. The opposite is true during the summer solstice. As can be seen, the more horizontally we move a string facing east (or west), the more the received flux increases. If we want to use a solar filament in the complement of latitude 49' L to heat the air during the winter solstice. The optimal inclination angle for it is 64.5 in the south direction. This angle of inclination at equinoxes also generates sufficient heat. But if this set is used to provide hot water in summer, the inclination angle is approximately 17.5 more effective. For one reason, we have not included the reflected radiation from the surrounding plain to the range. In addition, the flux predicted by the equation indicates that the pattern of sunshine around the solar noon is symmetrical. The amount of sun exposure in the morning and afternoon is usually different. As a result, east-facing and west-facing panels often do not receive the same amount of sunlight. A closer examination requires empirical information in which daily sun exposure is considered on a curved and sloping surface.

Panel string size

The size of a string is determined by factors such as environmental conditions, the need for heating, the efficiency of the string and the amount of sun exposure. Suppose, for example, the daily heat requirement of a house in the cold season is 100 kw-hr / day (or approximately 3.4×10^5 Btu / day) and the amount of sun exposure per day on the 4kw-hr / m^2 –day set Also assume that the area of each panel is 1.5 square meters, their range is 50% and one third of the heating is obtained with auxiliary heaters. Therefore, the amount of solar heating required is 66.7 kw-hr / day. Because the efficiency of the string is 50%.

Series and parallel strings

Each solar filament consists of thermal panels arranged in series, parallel, or a combination of the two. The output temperature of a large filament is not higher than the temperature that a single collector can produce. However, a filament with n panels has the potential to collect n times the amount of heat that can be obtained from a single panel.

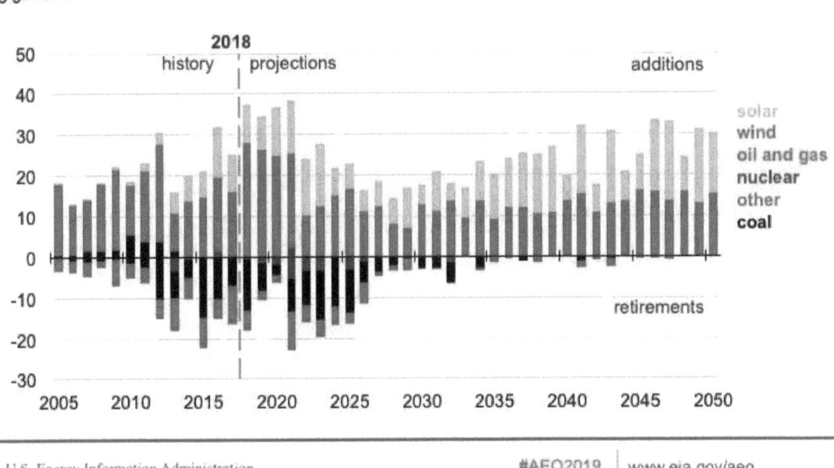

Figure 17. EIA Expects Renewable Growth to Slow

To collect this heat, the amount of fluid flux applied to the set n must be equal to. In a series string, the output of one panel is connected directly to the input of the next panel, so the increased flux must pass through all the panels of a string. As the fluid velocity increases, its resistance to flux increases. In addition, the longer the total length of the pipe through which the fluid passes, the greater. Therefore, a long series of series shows a high resistance to the flow of the transmitting fluid. To maintain fluid flow, the pumps must produce a high pressure so that the pressure at the inlet is much higher than the pressure at the outlet. This creates strain on the pumps and panels of a string. In addition, not all panels in a series work the same way. Panels that are closer to the inlet work at lower temperatures and are therefore more efficient. The opposite is true for panels that are closer to the fluid outlet.

In a parallel string, the inputs of each panel are connected to a common power line, and the outputs are similarly connected to a common drain pipe. Although making a parallel string is more difficult than a series string, it has less resistance to fluid flux. In addition, if the total input flow to the complex is evenly divided between the individual panels, the performance characteristics of the string can be easily obtained from the properties of a single panel. Increasing the efficiency and temperature of a parallel n-panel string is similar to increasing the efficiency and temperature of a single panel, but the flow rate and useful heat collected will be n. In practice, to facilitate the efficiency of the whole system, it can be analyzed in a similar way to that used for a single panel. Yield curves η, that is, curves in terms $\frac{(T_{fi}-T_a)}{F}$, can be plotted for a string. In addition, the outlet temperature can be determined after the flow rate is determined. If a controller is used, the flow rate can be adjusted so that the working temperature of the string is equal to the desired value.

Tube losses

Exhaust pipes, which transfer hot transfer fluids from one side of the filament to the other, waste some of their heat energy and return it to a cooler environment. Because ambient air acts as a heat storage source at Ta temperature, the heat transfer process can be approximated by using a single-current heat exchanger. If we apply the heat exchanger equation to an outlet pipe that operates the fluid at a high temperature (such as TN) from a string to a storage tank, we find that the temperature of the fluid reaching the tank is equal to:

$$(T_{Thermal} = T_a + T_H - T_a)\, exp\left(-\bar{u}_L L / m C_f\right)$$

$\bar{u}_L$ The total coefficient per unit length of the pipe for heat transfer from the fluid to the ambient air and L is the length of the pipe, the heat loss of the pipe is equal to:

$$Q_{Pipe} = mc_f(T_H - T_a)[1 - exp(-\bar{u}L/mc_f)]$$

The product of L H '= $L\bar{u}_L$ will be low for short and fully insulated pipes. In this situation we find that TH = storage is T and the loss of pipe is Q. The effect of pipe losses on small and large assemblies is significant.

Heat exchangers

The main use of solar heating is to provide hot water. Milk water can be heated directly with a string of solar panels. It can be consumed directly or, if higher temperatures are needed, it can be heated further with an auxiliary heater. However, in many applications it is necessary to transfer heat from a conveying fluid heated by the sun to a cold-water source. For example, the transfer fluid can be water to which ice powder has been added to prevent the set from freezing and corroding. The transfer fluid flows in a closed circuit (cycle) and the heat are transferred to the pipe connected to the valve through a heat exchanger with opposite flow. It can also transfer the heat to the water storage tank through a single current exchanger.

A) An open circuit system in which tap water is first partially heated by solar energy and then heated by an auxiliary heater.

B) A closed-circuit system in which heat is transferred from a conveying fluid heated by sunlight to water with opposite currents.

C) A closed-circuit system in which heat is transferred to a water storage tank by a single-current converter.

As we know, the rate of heat transfer from a warmer fluid to a colder fluid in a reverse flow heat exchanger is:

$$Q = \bar{u}_L LA$$

Where $\bar{u}_L$ the heat transfer coefficient and L are the length of the converter. The logarithm of the mean temperature difference between hot and cold signals is:

$$A = \frac{(T_H - T'_H) - (T_C - T'_C)}{ln[(T_H - T'_H)/(T_C - T'_C)]}$$

T_H and T_C are the hot (inlet) and cold (outlet) temperatures of the primary fluid (heated by sunlight) and T'_H and T'_C are the hot (outlet) and cold (inlet) temperatures of the secondary fluid (e.g. water). The rate of heat transfer $\dot{Q} = \dot{m}_f C_f (T_H - T_C)$ is related to the relation: or equivalent to the opposite relation:

$$\dot{Q} = \dot{m}'_f C'_f (T'_H - T'_C)$$

Solar heat from the filament can also be stored in a large tank using a single-current heat exchanger. The transfer fluid passes through a heat exchanger which is itself immersed in a storage medium with a temperature of T_B. If the temperature of the inlet fluid is T_H and $\dot{m}_f$ its flow rate is:

$$T_C = T_B + (T_H - T_B) \, exp\left(-\bar{u}_L L / \dot{m}_f C_f\right)$$

Comes out:

$$Q_{Store} = \dot{m}_f C_f (T_H - T_B)\left[1 - exp\left(-\bar{u}_L L / \dot{m}_f C_f\right)\right]$$

It is stored in the environment.

Transducers can exchange heat between two liquids or between a liquid and air. For example, in systems that heat the space by blowing air, heat can be transferred to the blown air stream in two ways, first through the liquid inside the heat storage tank and secondly through the opposite flow of a convection fluid by sunlight. It's hot. The more efficient the converter, the greater the amount of heat transferred. In addition, the air is heated to a higher temperature and the transfer fluid is cooled more efficiently.

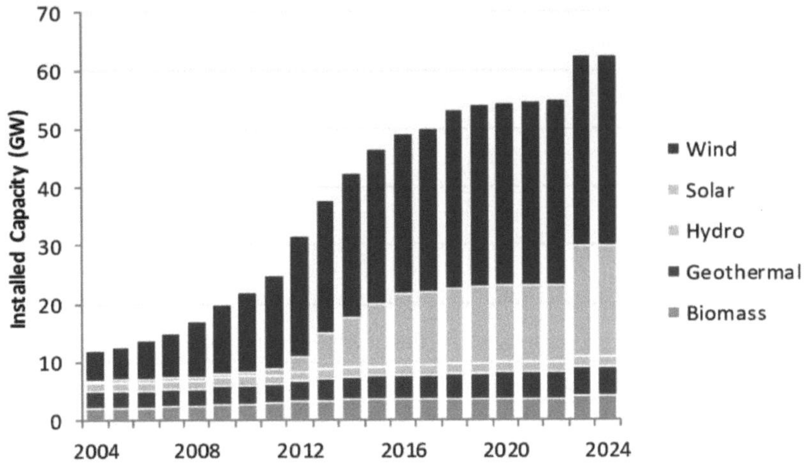

Data Source: WECC 2024 Common Case

Figure 18. Historical & projected growth of renewables

Storage

In a solar heating system, measures are usually taken to store heat. The energy obtained can be stored during times when the sunlight is intense and used during times when the sun is hidden. When heat is stored in an intermediate, the intermediate energy increases. This increase can also be in the form of potential, but in this case the molecular structure changes, such as a chemical change, a phase change (such as melting or evaporation). When the heat is increased, it only increases the temperature in the environment. We say that thermal energy is stored as tangible heat. As long as there is no phase change, the temperature increase is almost proportional to the stored heat and is inversely proportional to the mass, and the temperature change can be written as follows:

$$(\text{Tangible heat storage}) \qquad \Delta T = Q_S/m_S C_S$$

Where Q_S is the stored heat, C_S and m_S are the specific heat and mass of the storage medium, respectively. When tangible heat is taken from the environment, the temperature drops according to Equation. An important parameter in a thermal storage system is the heat stored per unit volume. To store tangible heat, Equation can be written as follows:

$$\frac{Q_S}{V_S} = \frac{m_S}{V_S} C_S \Delta T = P_S C_S \Delta T$$

In some applications, it is desirable to keep the temperature of the storage medium to a minimum. According to Equation 2-9, this change can be reduced by increasing the mass of the storage medium. But a more effective way to stabilize the temperature is to use latent heat instead of storing tangible heat. When we heat a solid, its temperature begins to rise to a melting point of T_m. When we give it more heat, its temperature remains constant at T_m, but it undergoes a phase change and turns into a liquid, and the temperature does not rise until the melting operation is complete. The heat that is absorbed by the environment per unit mass and stored during the phase change is called latent heat l. The latent heat per unit volume is from:

$$(\text{Latent heat}) \frac{Q_S}{V_S} P_S l$$

PS is the average density of the medium. For example, the latent heat of ice J / kg 5 is $1=5/35\times10^5$ J/Kg which is equal to 0.093 kw-hr / kg. If we assume that the density of the water-ice environment is approximately equal to the density of water, i.e. $P_S = 1000$ kg / m³, we conclude from Equation that:

$$\frac{Q_S}{V_S} = (1000)\,(0.093) = 93 \mathrm{Kw\text{-}hr/m^3}$$

Comparing this value with water in Equation, we find that if we let the water temperature change 80° C with the same tangible temperature we can achieve the same storage density, latent heat storage is not accompanied by any change in storage temperature.

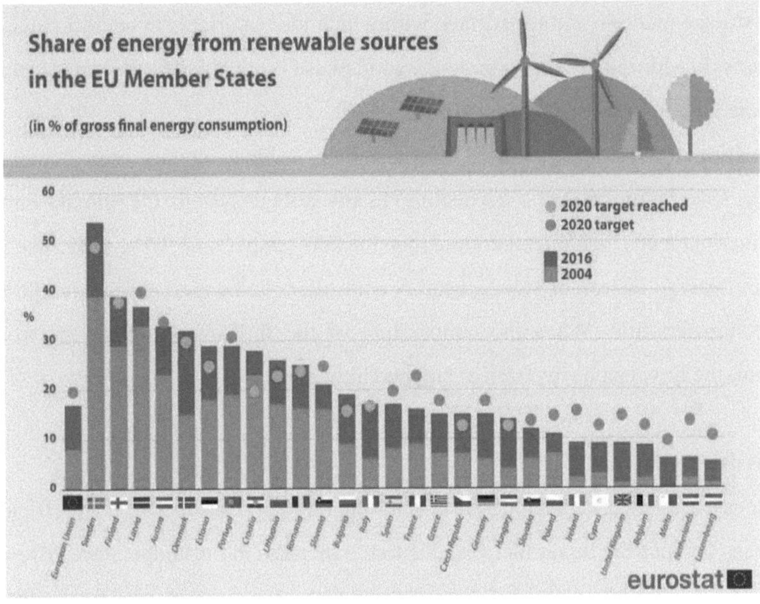

Figure 19. Terra Form Power: An Undervalued Renewable Energy Growth Story
(NASDAQ: TERP)

The problem with storing latent heat in ice water is that the melting point is too low to be useful. Thus, the goal is to find a solid whose melting temperature is lower than what the solar system can produce and at the same time its temperature is higher than the level required for heating. The melting point of many solids at room temperature is so high that they are not useful in conventional solar storage applications, although the phase change temperature of some hydrated salts is low and suitable for storing solar energy.

In fact, certain mixtures of salts melt at lower temperatures than their constituents. Such mixtures are called ethical mixtures. The melting temperature of such mixtures can be somewhat reduced by choosing the right mixing ratios. Storing thermal energy in the form of latent heat has certain advantages over storing it in tangible heat. This storage takes place in a narrower range of temperature and the energy per unit volume is significantly higher. Of course, there are some technical issues. Chemical spoilage due to the storage medium and the surface with which the material is in contact can cause problems. In addition, heat transfer to the solid phase is not always efficient, especially if the mediator of a thermal conductor is weak.

The melting of the storage medium must also be possible at a temperature compatible with the specific function of solar heating. In any storage system, the primary transfer fluid must enter at a higher temperature than the tank temperature. Otherwise. The fluid will emit energy instead of storing heat. A controller can be used to regulate the flow of the transfer fluid. When the temperature of the fluid drops below the storage medium, the flow can be diverted or stopped altogether.

Sun-assisted systems

As discussed, as a working temperature of a solar thermal panel increases, its efficiency decreases because the losses increase due to the increase in the temperature difference between the absorber plate and the environment. Assuming a moderate carpanel temperature $\bar{T} \cong \frac{T_{f.e}+T_{f.i}}{2}$, it should be as small as possible to maximize efficiency $\bar{T} - T_a$. In order to be low, the temperature of the fluid must be as low $\bar{T}$ as possible when entering and leaving. Consider the moment when the inlet temperature of T_{fi} is constant (for example, in a hot water supply system T_{fi} is stabilized by the temperature of the cold-water source, while in a space heating system the value of T_{fi} by the temperature The air inside the room is fixed.) Therefore, in order to reduce $\bar{T}$ and increase the efficiency, we need to reduce the T_{fi} outlet temperature by increasing the flow rate of the transfer fluid. If we see that this temperature is too low to be useful, it can be increased by an auxiliary heater to an acceptable level. The increased efficiency of the assembly work means that fewer panels are needed. In fact, in harsh environmental conditions, its passive temperature may be lower than the level required to generate useful heat. In this situation, the set will not be able to work at all without an auxiliary system.

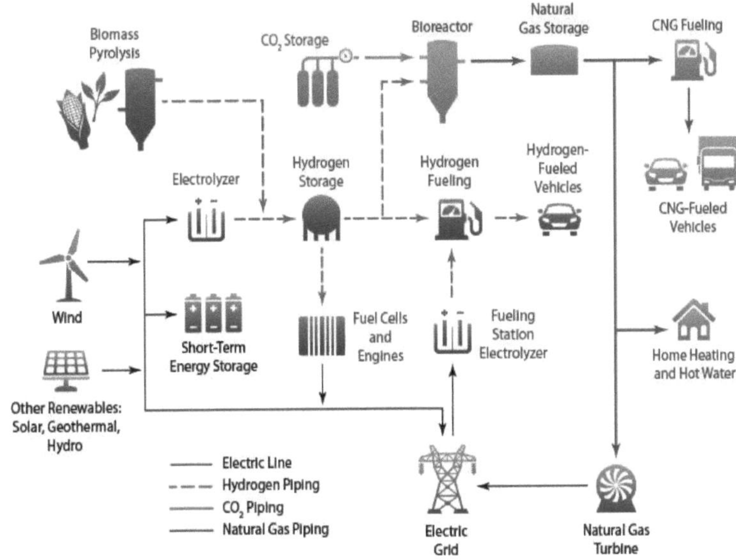

Figure 20. Renewable Electrolysis | Hydrogen and Fuel Cells

Example: A collector set has a 60% efficiency when it brings the cold-water temperature of milk from 10 °C (50 °F) to 45 °C (110 °F) when the same water temperature rises to 60 °C (140 °F). Bring the string efficiency to 40%. The system is designed to provide hot water at a temperature of 60 °C. Compare the string size for a system without an auxiliary heater with a system in which an auxiliary heater is used to raise the water temperature from 45 °C to 60 °C. Suppose both water systems are equally hot they do. Suppose it is necessary to supply 100 units of heating power by each system.

At 40% efficiency, the system without auxiliary heater should receive 250 units of radiant power. The system with auxiliary heater raises the temperature to 35 degrees Celsius and the auxiliary heater heats 15 degrees Celsius. Because the set with the auxiliary heater has an efficiency of 60%. The received radiant power should be 1116.4

units. As a result, the filament without auxiliary heater heats more than twice the number of panels of a filament with its auxiliary heater by 15 degrees Celsius. Because the set, together with the auxiliary heater, has an efficiency of 60%. The received radiant power should be 1116.4 units. As a result, a string without an auxiliary heater requires more than twice the number of panels in a string with an auxiliary heater. Auxiliary heater is provided in the system.

Solar-assisted heat pump

In solar (air) heating applications that operate without the aid of the sun, the working temperature of the collector is limited by the fact that T_{fi} cannot be lower than the temperature of the space to be heated, for example, if a conveying fluid Just as air heated by sunlight transfers its heat to a room, its temperature cannot be colder than room temperature when it returns to the collector. In fact, this temperature (T_{fi}) is probably significantly higher than the ambient temperature to be heated. This limits the efficiency of the field. Using a device called a heat pump, heat energy can be taken from the transfer fluid and the equipment can be returned at a temperature lower than the desired space temperature. This will significantly increase the efficiency of the field.

If the filament is equipped with a heat pump, it can actually operate at a temperature that is lower than the temperature of the space to be heated. Heat can be transferred from a colder area (i.e., from the filament) to a warmer area (heated space). To understand how heat flows from a cold to a hot environment, it is necessary to consider the second law of thermodynamics.

The law states that it is impossible to build a device that does not produce any effect of heat transfer from a cold storage source to a hot source in its working cycle. Here's what's important about a heat pump. The second rule prevents heat from flowing from a cold zone to a hot zone unless there is another effect, such as the use of labor. Therefore, a device can be built that transfers heat from a colder area to a warmer area when given the job. An example is an air conditioner. On a hot summer day, the temperature inside a room may be 20 ° C and the outside air may be 30 ° C. If the air conditioner does not work, heat will flow out of the room and the room temperature will gradually rise. When the air conditioner is turned on and given electrical work, it takes the heat from the colder room and returns it to the warmer air, which keeps the room temperature.

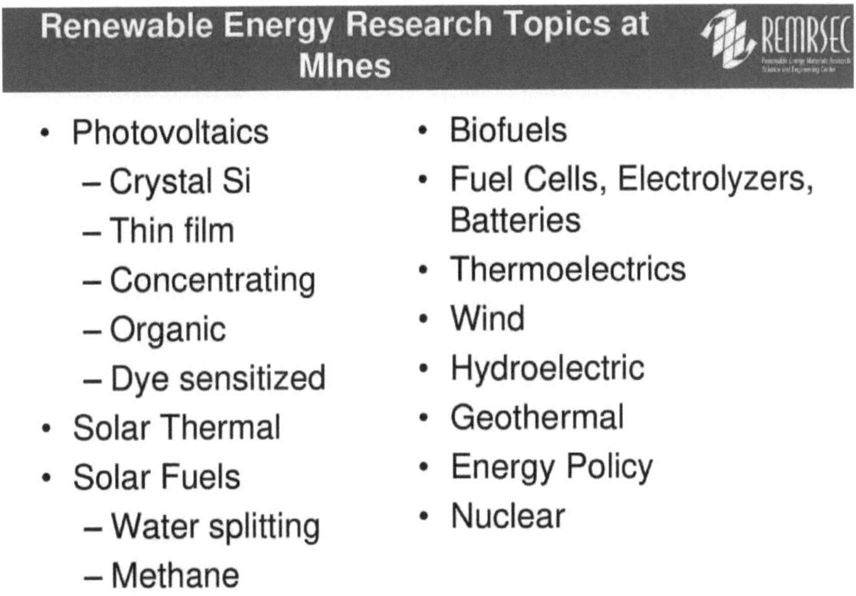

Figure 21. Research for Renewable Energy Applications at Mines and NREL

Now consider a cold day where the outside air is 10 ° C and the room temperature is 20 ° C. If we do not heat the air in the room, it will gradually cool down and eventually reach c10. Suppose we have installed an air conditioner next to a window but it is on the outside of the room, in this case the device takes the heat out of the colder environment and transfers it into a warmer room and therefore the room temperature in 20°C remains constant. The device that generates heat in this case is called a heat pump. The heating given to the room by a heat pump is equal to the heat taken out of the air plus the equivalent of the work given to the device. Hence, we can write:

$$Q_H = Q_c + W$$

Q_H is heated by the hot heat source, Q_C is heated by the cold heat source and W is applied to the heat pump. The efficiency coefficient (COP) of the heat pump is determined as follows:

$$(\text{Heat pump} = COP = \frac{Q_H}{W})$$

The higher the efficiency, the more efficient the heat pump. The minimum possible value for the COP is unit 1. This value corresponds to a time when work (for example, electrical energy) is converted directly to heat and no heat is taken from the cold source. We show how to place a solar-assisted heat pump in.

A heat pump takes heat from a low-temperature fluid and returns it to a high-temperature environment. In fact, the job of the heat pump is to take heat from it and cool it before the transfer fluid returns to the heating panel. Hence, the heating panel operates at lower temperatures and loses heat. It becomes less environmentally friendly. The heat pump raises the heat to a level that is suitable for special applications. The total heat given is equivalent to the heat taken from the transfer fluid plus the equivalent of the required work of the heat pump. The lower the amount of work required by the heat pump, the higher the COP efficiency coefficient. In the equation we have new work:

$$(\text{Heat pump}) \; COP_{real} \leq COP_{Carno} = \frac{T_H}{T_H - T_C}$$

T_C and T_H are in degrees Kelvin. The heat pump operates at a new level if it uses a reversible thermodynamic cycle called the Carnot cycle. To use a reversible cycle called the Carnot cycle. The Carnot heat pump is an ideal example and cannot be built for operation. Of course, the Carnot limit provides a small measure of the efficiency of the heat pump. We see from Equation that as the temperature difference between the sources increases, the efficiency of the heat pump decreases. Although a heat pump can be used to obtain heat energy from outside air at $10°$ C and enter a room with a temperature of $21°$C, it does so by lowering the outside air temperature and approaching the freezing point. Its efficiency decreases. When the temperature is below freezing point, the solar filament can be used to bring the temperature to a level that is acceptable for work. Interestingly, there is an optimum operating temperature for a solar-assisted heat pump. If the filament operates at a low temperature, its efficiency is high, but the COP efficiency coefficient of the heat pump increases, but the efficiency of the filament approaches zero.

Here it is appropriate to use a temperature controller to establish an optimal outlet temperature of the filament. A system that is powered by the sun and uses a heat pump. The set transfers heat to the transfer fluid and stores it in a special hot water tank. When heat is needed but the sun is shining, controller 1 gradually fills the hot water tank with hot transfer fluid at a predetermined temperature, T_{fi}.

This process stops when the cold-water storage tank is empty and the hot water tank is full. When heat is required, the controller draws fluid from the hot tank and returns it to the cold-water tank. The heat is taken up by the heat pump and the transfer fluid loses heat before returning to the tank containing the cold fluid. The controller adjusts 2 fluxes to set the right temperature for the heat pump cold source. In general, the heat pump acts as a superior heat exchanger that transfers heat from the solar system's transfer fluid to the air stream. The heat pump allows heat to be transferred from the colder transfer fluid to a warmer air stream. As a result, the temperature of the air given to the room is actually higher than the output temperature of the solar filament. This state of heat transfer can only occur if given some work. The greater the temperature difference between the heat sources, the more work is required.

Radiant coolers

We conclude this chapter with a brief discussion of the use of solar thermal panels for radiant cooling. As mentioned in the previous section, an air conditioner is needed to take heat from a cold source and deliver it to a hot source. The higher the temperature of the hot source, the lower the efficiency of the cooling process, and the hotter the outlets (i.e., the hot source) of the air conditioner. It is becoming more and more difficult to cool the room. If we immerse the condenser coil tubes (i.e. the tubes that carry heat to a warmer environment) in a colder fluid, the efficiency of the cooling process will increase. Suppose that at night, when the ambient temperature is relatively low, a transmitting fluid flows through the solar panels. On clear nights, heat loss caused by radiant cooling can lower the temperature of the fluid from the ambient air. Colder fluid can be stored at night and, if necessary, passed through the front of the condenser coil condenser tubes. This will increase the efficiency of the air conditioning process.

Unfortunately, although the design of a solar thermal panel is better, it is less efficient as a radiant cooler at night because good solar panels are designed to minimize heat loss, such as a solar panel on the screen. Coated with a selective coating, it does not absorb heat more effectively. A good radiant cooler requires a fully radiant surface. In addition, it is necessary to remove the cap at night to increase convective losses.

So far we have dealt with the production of heat from solar energy. Most of the energy consumed in an industrial society is in non-heating applications, which fall into a category called labor. For example, electricity is a form of energy that allows electric motors to work. Similarly, the chemical energy stored in fossil fuels is used in internal combustion engines and steam engines to get the job done.

Chapter VI

Sunbathing on the Ground

The amount of solar energy on the Earth's surface is significantly less than the amount of energy that reaches the Earth's atmosphere. The rate of reduction of solar energy when entering the Earth's surface is mainly determined by the light state of the Earth's atmosphere. As we will see later, the compositional components of the atmosphere affect solar radiation by two processes. In certain areas of the spectrum, the sun's energy is mainly scattered, while in other areas most of it is absorbed.

Therefore, the spectral composition of the sun on the surface of the earth is significantly different from that of the 5760 K solar blackbody curve. It is also important to note that ground-level sunbathing can no longer be equated with a one-way beam. This was true of radiation reaching the upper atmosphere. Some of the radiation scattered by the atmosphere reaches the earth in the form of scattered radiation. Scattered radiation is a component that travels in different directions. Therefore, total solar radiation on the Earth's surface consists of a direct, one-way component that creates atmospheric scattering in quantitative terms because of how it changes and to understand the conversion of solar energy after passing through the atmosphere, we provide some basics of atmospheric physics.

An atmospheric model

Atmospheric state can be detected to some extent by thermodynamic variables such as temperature T, density P, pressure P and chemical composition. These parameters change according to the spatial and temporal position in the atmosphere. Because this change is relatively unpredictable. It is very difficult to make theoretical estimates about sunbathing on the surface. In order to obtain some theoretical conclusions, it is necessary to make a few simplistic definitions of atmospheric structure. First, we assume that the atmosphere is thin enough to be considered flat relative to the earth's radius. As we will see, the effective altitude of the atmosphere is approximately 8 km, which is very small compared to the radius of the earth (R = 6371 km).

Therefore, it is a good approximation, except probably around sunrise and sunset, when the sunshine on the earth's surface is so low that it can be ignored. Therefore, atmospheric curvature is of little importance in most solar energy applications. In the second approximation used here, it is assumed that atmospheric parameters change with only one characteristic, the height Z. That is, all atmospheric parameters can be displayed in terms of vertical profiles such as P = P (Z), T = T (Z) and P = P (Z). The accuracy of this approximation has not been proven, especially when there are scattered clouds in the sky. A thin atmosphere whose composition changes only with altitude. The atmosphere is called layered.

Absorption and scattering of solar radiation by atmospheric components

Atmospheric components, whether molecules such as n_2, o_2, Co_2, and H_2o, or ozone, and larger particles such as droplets of fog, soot, or dust, can affect radiation by absorption or dispersion. In the absorption process, the radiated energy is converted into another form of energy, usually heat. Part of the absorbed fraction is determined to some extent by the cross-sectional area of the mass absorption.

This parameter varies from one molecule to another and $\sigma^a(\lambda)$ also depends on the wavelength of the radiation. As we will see, the molecules n_2, o_2 are not significantly absorbed in the solar spectrum, on the other hand Co_2 and H_2o in Selected areas are highly absorbed by the infrared region of the solar spectrum. These areas are called specific absorption drivers. In the ultraviolet region of the solar spectrum, ozone-absorbing bands form in the stratosphere. Dispersion is more complex than adsorption.

Like the absorption process, a fraction of the radiated energy is lost. This value $\sigma^a(\lambda)$ is determined by the cross-sectional area of the mass scattering component. Scattering, unlike adsorption, does not convert radiant energy into heat but redistributes it in other parts of space. Atmospheric scattering of solar energy on a clear day is mainly caused by oxygen and hydrogen. The theory states that according to Riley's law, the scattering of solar energy by air molecules changes slowly relative to the wavelength.

$$\sigma^s(\lambda) = \frac{C}{\lambda 4}$$

Where c is a parameter that depends slightly on the wavelength. According to Riley's law, short wavelengths such as ultraviolet, purple and blue are more scattered than red and infrared wavelengths. Hence, ordinary air has a wide dispersion in the visible region of the spectrum.

Especially for the performance of the blue and purple constituents that justify the sky-blue color. Some particles cause scattering that has preferential forward and backward directions (Rayleigh scattering) while other radiation particles scatter more uniformly. Particulate matter in the atmosphere, such as dust, soot, and fog, is exposed to radiation. They scatter more complex than what is predicted by Rila's law. The red color of the sky at sunset is the result of the scattering of radiation by dust particles adjacent to the earth's surface. Radiation that is not subjected to absorption and scattering processes is called the direct or attenuated component. This parameter and how it affects the direction of radiation will be described in detail later.

Direct sunlight

We assume that the atmosphere of the executive earth is uniformly mixed and also has a density profile $P = P(Z)$. So far, we have defined a point in the atmosphere by altitude with a height variable (Z) whose value varies from zero at ground level to infinitely above the atmosphere. Because solar radiation reaches the top of the atmosphere and is scattered from there to the earth's surface, it is better to present a variable such as s that measures the point depth inside the atmosphere relative to the top of the atmosphere. This variable varies from $S = 0$ in the upper atmosphere to the surface $S = \infty$. The increase in height and depth are related according to the relation $dz = ds$. Suppose a beam of parallel rays of the sun with spectral intensity and at an angle to the apex z shines on an extremely thin layer of atmosphere ds (at depth s). The fractional change in light intensity when the beam This appears from the bottom of the layer, which can be expressed as follows:

$$\frac{dI_\lambda}{I_\lambda} = -P(S)6(\lambda)d, I$$

Where P (s) and $\sigma(\lambda)$ the mass density and $[\sigma^{a}(\lambda)+\sigma^{s}(\lambda)]$total cross-sectional area of the layer, respectively. The oblique thickness of the beam is obtained from the following formula $(d\lambda)$:

$$dl = \frac{ds}{\mu_0}$$

Which is $\mu_0 = CZ$. It is better to simplify equation 3-8 and write:

$$\frac{di}{I_\lambda} = k_\lambda(s)\frac{ds}{\mu_0}$$

Which $k_\lambda(s) = P(s)6(\lambda)$is called the spectral attenuation or extinction coefficient. This coefficient varies according to the height of the layer and the wavelength of the radiation. To find the amount of attenuation created by the whole atmosphere, we say from Equation for s = 0 to $s = \infty$the integral. This operation is performed due to the fact that at s = 0 (above the atmosphere) the spectral distribution is equivalent to the constant solar distribution. Thus, the direct or attenuated spectral intensity at the earth's surface is equal to:

$$= s_\lambda e^{-t_\lambda/\mu_0} \, I_\lambda \, \text{Direct}$$

Which s_λ refers to the intensity of the fixed spectrum of the sun and $t_\lambda = \int_\lambda^\infty k_\lambda(s)ds$ the light spectral thickness of the atmosphere related to the beam of radiation λ with wavelength is called. This direct intensity essentially remains unidirectional or species-like while passing through the atmosphere. It decreases with increasing inclination to the radiation. The exponential dependence on Cz indicates that the existing solar flux decreases significantly as the time of sunrise and sunset approaches.

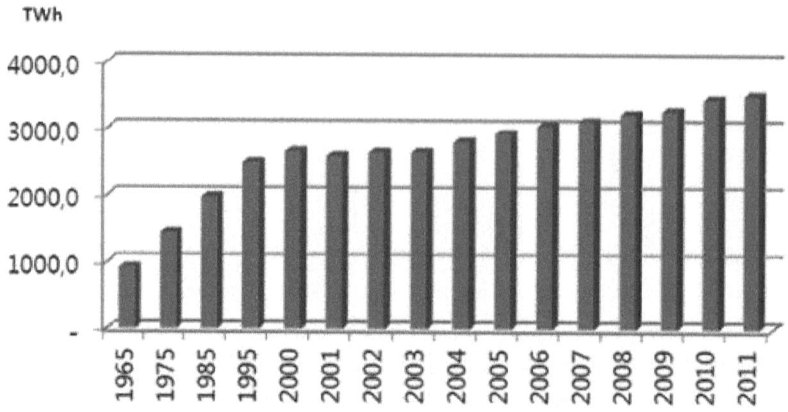

Figure 22. Technology and Applications

Broadcast flux

The total solar flux that shines on the sea surface consists of a one-way beam of direct flux plus a component of the scattering flux. The diffusion component consists of radiation or reflected from the components of the atmosphere, as well as radiation reflected from the ground beneath it. There is nothing for it. Here, instead of trying to find accurate solutions to the diffusion flux, we present an analysis to estimate it and also to show how factors affect the environment. So far, we have not been particularly careful about the distinction between the word's intensity and flux. Although this distinction is not necessary for unidirectional radiation, it is really necessary for radiation. The term intensity is used to describe the distribution of directional radiation. In particular, it provides information such as: how much radiant power is propagated per unit area in different directions around a point in space. We specify a specific direction as a vector Ω. For a flat layered atmosphere, we can write the spectral intensity $I_\lambda(Z,\Omega)$. This figure suggests that the spectral intensity is a function of both the height and the direction of the case. Although the direct solar component is essentially unidirectional, the scattering component is not simply composed of identical radiation. Note that in the description of the strict function, no interrupting surface is mentioned.

On the other hand, the concept of flux only makes sense about a certain level. Consider a small flat surface, with a vertical inward direction, that is, $\hat{n}$ with a vector that is determined in the vertical direction into the mentioned surfaces. The spectral flux received by this surface is as follows:

$$F_\lambda(Z,\hat{n}) = \int_{(n,\hat{\Omega} \geq 0)} I_\lambda(Z.\Omega)\hat{n}.\Omega d\Omega$$

Which $\hat{n}.\Omega$ is the product of the cosine $\Omega, \hat{n}$ of an angle. The index above the integral reminds us that $\hat{n}.\Omega > 0$ integration takes place on the spatial angles in which it is located. This assures us that the only radiation transmitted in the hemisphere to the surface is listed in Equation. Equation shows that the flux received by a surface, even for diffuse radiation, generally depends on the direction of the surface and its position in the atmosphere. In the case of cheek radiation, it can be proved that Equation is simplified as follows:

$$F_\lambda^{Light} = F_\lambda^{Vertical}\hat{n}.\Omega_\circ = F_\lambda^{Vertical}G\theta$$

The angle of inclination θ between the beams of the line perpendicular to the surface. $F_\lambda^{عمودی}$ A flux is a ray that shines on a surface whose direction is perpendicular to the beam. For matched radiation, one can take integrals from Equation 5-8 to obtain the following formula:

$$F_\lambda^{Similar} = I_\lambda^{(Similar)} \int_{(n,\hat{\Omega} \geq 0)} \hat{n}.\Omega d\Omega = \Pi I_\lambda^{Similar}$$

In this case, as expected, the reception is independent of the surface orientation. The diffused radiation is generally not uniform and the received flux depends to some extent on the surface orientation. This dependence is similar to what is seen in the case of direct flux

Approximate equations for total solar flux:

A complete and comprehensive discussion of total intensity (direct intensity plus scattering) for sunbathing on Earth requires some of the most advanced and sophisticated knowledge familiar to mathematical physicists. To find the intensity we have to solve a differential integral equation. It is known as the radiation transfer equation. Even for a flat layered atmosphere, it is very difficult to reach solutions. This problem increases with the complexity of the molecular dispersion process. Solving the intensity function becomes more and more complicated as a result of reflection from the ground. Even an approximate solution $I(Z, \Omega)$ of the radiation transfer equation is beyond the scope of this text. To read more, we make a few simplistic assumptions:

> ➢ The diffusion component is much smaller than the direct flux.
> ➢ The downward diffusion component that shines on a horizontal surface is identical.
> ➢ It is the lower surface of Lambertian, that is, it reflects radiation according to Lambertian law.

The Lamberti surface is the surface at which the reflected radiation is always the same regardless of the nature of the irradiated radiation. Therefore, the atmospheric flux is assumed to be completely diffuse or homogeneous. Using these assumptions, we see that the diffusion flux is composed of an identical downward flux, $F^{\downarrow(partial)}$ and $F^{\downarrow(partial)}$ an identical upward flux. In addition, the total diffusion flux that reaches a diagonal surface depends on its inclination relative to the vertical line, not its lateral angle. It can be proved that the total diffusion flux that reaches ϕ a diagonal surface for upward fluxes and the bottom of the match is:

$$F^{\uparrow(partial)} \quad F^{+}\left[\frac{1-Cs\Delta}{s}\right] + \left[\frac{1+Cs\Delta}{s}\right] F^{(partial)}$$

So that the total received flux (direct and diffuse) is equal to:

$$\underset{F}{(Direct)} + \underset{F}{(partial)} = F$$

$$\underset{F}{\uparrow(partial)} \underset{F}{+} \left[\frac{1-C\Delta}{2}\right] + \left[\frac{1+C\Delta}{2}\right] F = s\mu e^{-j/\mu 0}$$

The diffusion flux ($\Delta = 0$) is completely horizontal $\underset{F}{\downarrow(partial)}$ for one level. While this flux ($\Delta = 90°$) is equal to $\underset{F}{\downarrow(partial)} + \underset{F}{\uparrow(partial)} = \underset{F}{(partial)}$ for a vertical surface.

After the two fluxes $\underset{F}{\uparrow(partial)}$ and $\underset{F}{\uparrow(partial)}$ were determined. Calculate the total flux on a sloping surface from Equation 5-9. Using some simplifying approximations, we can convert the radiative transfer equation to a pair of simple ordinary differential equations for up and down fluxes. This equation includes three environmental parameters: light atmospheric thickness J, reflection coefficient with A subtraction$\bar{w}_o$, and the reflectivity (R) of the lower surface. The last two parameters are defined as follows:

$$\bar{w}_o = \frac{J^{(s)}}{J}, \qquad R = \frac{F\uparrow}{F\downarrow}$$

$J^{(s)}$ The light thickness of a ray of scattering and $F^{\uparrow}$ and $F^{\downarrow}$ the downward and upward fluxes on the earth's surface, respectively, in a fully diffused atmosphere (i.e., an atmosphere in which there is no absorption at all) $\bar{w}_o$ is $= 1$, while for a completely absorbing atmosphere (Atmosphere in which there is no dispersion) $\bar{w}$ is $= 0$. For a completely covered area of trees and plants, it is R <0.2, while for a snow-covered area, it is R $= 0.5$. Of course, the parameters J and R depend on the wavelength, but we will use values that are average values across the spectrum. The upward and downward diffusion components on the ground surface, according to these parameters are as follows:

$$F^{\downarrow(partial)} = s\mu_0 \left[\frac{1}{1+G}\left(Ge^{\delta+J} + e^{\delta+J}\right) - e^{-J/\mu_0} \right]$$

$$F^{\uparrow(partial)} = Rs\mu_0 \left[\frac{1}{1+G}\left(Ge^{\delta+J} + e^{\delta+J}\right) \right]$$

$$G = -\left[\frac{\delta^- + A - BR}{\delta^+ + A - BR}\right] e^{(\delta^- - \delta^+)J}$$

$$\delta^{\pm} = \frac{1}{2}(C-A)^+ - \frac{1}{2}[(C+A)^2 - 4BD]^{1/2}$$

$$A = \left(\frac{2-\bar{w}_0}{2\mu_0}\right), \qquad B = \bar{w}_0, \qquad C = (2-\bar{w}_0), \qquad D = \frac{\bar{w}_0}{2\mu_0}$$

It is easily provable that whenever there is no dispersion. ($\bar{w}_o \rightarrow 0$) The downward scattering flux really depends on R (via the G function) because the reflected radiation can be redistributed to the earth by the atmosphere.

Measurement of sun exposure on the ground

Although the previous analysis provides insight into the determinants of sun exposure on Earth, it is better to have empirical data to compare the theory with. Such data are also important in determining the efficiency of solar energy wires. The data obtained over several chapters provide statistical information on the availability of applications of certain solar energy systems in specific geographical areas. Most energy meters are in one of two categories:

Photoelectric and radiation measuring devices The first group includes devices with receiver or sensor parts whose electrical characteristics change when there is solar radiation. For example, it generates voltage when sunlight shines on photovoltaic devices such as silicon and silicon cells. The short-circuit current of these devices is used to measure the intensity of the irradiated radiation. Optical conductive detectors, such as cadmium sulfide or cadmium silica, change resistance in response to electromagnetic radiation when these detectors are connected to a battery. In the circuit, they become a criterion for measuring the level of intensity. In addition, there are devices with vacuum tubes known as incandescent bulbs, which are coated with special elements that emit electrons when light shines on them.

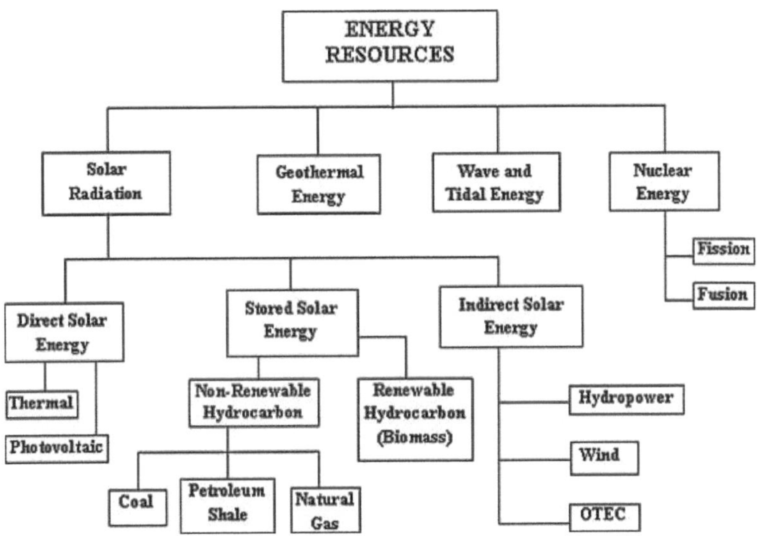

Figure 23. Strategies for renewable energy applications in the organization of Islamic conference (OIC)

Radiation intensity is used. Although photovoltaic transistor devices are durable, small, and inexpensive to make and are rarely affected by environmental conditions, they do have drawbacks. The first of these devices tends to be interrupted or saturated at high intensity levels. Second and most importantly, photoelectric devices do not provide a uniform spectral response across the solar spectrum. This means that equal amounts of solar energy received at different points in the spectrum produce different signals. Especially if a device is only sensitive to the visible spectrum.

This device will not detect the presence or changes of solar energy in the infrared. The second group includes radiation meters. These devices usually absorb radiation in a black absorber and use the heat generated to create a receiver. This change is measured and related to the levels of sun exposure. A radiation meter is a type of black and white pyranometer used in the United States. The receiver consists of two adjacent planes, one black and the other white (silver).

A thermal sensor (usually a thermocouple) is usually connected to each of these parts. The sensors generate a voltage signal that is directly proportional to the temperature difference between the black and white surfaces. When solar energy shines on the device, the black surface absorbs the color of the radiation and the white surfaces reflect the radiation. And as the ambient temperature remains, it gets warmer. The higher the level of sun exposure. The greater the temperature difference of the voltage signal, the larger it becomes.

Atmospheric heat flux

As the last article of this chapter, we examine the thermal radiation of atmospheric radiation on a surface. Although this radiation cannot be called thermal energy, but in any case, plays an important role in the performance of the steady state of solar thermal panels. The amount and spectral distribution of the irradiated radiation depends on the temperature and the ability of the object to absorb (emit). The atmosphere and its underlying region absorb solar radiation and emit heat radiation. These emitters are usually at a temperature of about 300 k. Most of the energy spectrum of a black body at this temperature is in the interval $2\mu m\lambda \leq 20\mu m$, and only a negligible fraction of solar energy reaches this wavelength. It is common to call solar energy short-wavelength radiation $0.3\mu m\lambda \leq 20\mu m$, while thermal energy is called long-wavelength radiation dt_λ. Consider a differential layer of light-thick barley at light depth t_λ. Consider a beam θ whose angle of μ view of the intensity of the beam can be shown as:

$$dt_\lambda = \frac{B_\lambda(T)}{\Pi}dt_\lambda/m \qquad (M = C\theta)$$

Which $B_\lambda(T)$ is a function of Planck and T is the temperature of the layer. This beam travels towards the ground and is attenuated by the lower layers of the atmosphere. The intensity of the beam when it reaches the ground is:

$$(dt_\lambda^{(Thermal)} = \{exp[-(J_\lambda - t_\lambda/\mu)]\}dI_\lambda^{(i)\,(Thermal)} = \frac{B_\lambda(T)}{\Pi}\{exp\,l\,[-(J_\lambda - t_\lambda)/\mu]\}\frac{dt_\lambda}{\mu}$$

Which $(J_\lambda - t_\lambda)$ is the optical thickness of the air between the emitter layer and the ground. The total intensity is obtained by adding the role of all layers of the atmosphere by integrating equation:

$$I_\lambda^{(Thermal)} = \frac{1}{\pi}\int_0^{t_\lambda} B_\lambda(T)\,exp[-(\tau_\lambda - t_\lambda)/\mu]\,dt_\lambda/\mu \qquad 0 \le \mu \le 1$$

Since T is a function of altitude, integration can only be done after determining the temperature profile. If we assume that the atmosphere is isothermal, then $T_0 = T$ can be integrated from Equation as follows:

$$I_\lambda^{(Thermal)} = \frac{1}{\pi}B_\lambda(T_\circ)\left(1 - e^{-T_\lambda/\mu}\right) \qquad 0 \le \mu \le 1$$

In addition, if the optical thickness of the atmosphere $2\mu m\lambda \le 20\mu m$ can be approximated by a constant figure equal to the average value of the spectral range (that is, $\bar{\tau}_\lambda = \tau$ then Equation can be written as follows:

$$I_\lambda^{(Thermal)} = \frac{1}{\pi}B_\lambda(T_\circ)\left(1 - e^{-T_\lambda/\mu}\right)$$

The downward heat flux that shines on the horizontal surface is obtained by using Equation and integrating on a hemisphere in the downward direction:

$$I_\lambda^{\downarrow(Thermal)} = 2\pi \int_0^1 I_\lambda \mu\, d\mu = 2B_\lambda(T_0) \int_0^1 (1 - e^{-J/\mu})\mu\, d\mu$$

$$= 2B_\lambda(T_0)\left[\frac{1}{2} - E_3(J)\right]$$

Which $E_3(J)$ is called the exponential integral of the third order. To find the total thermal radiation across the spectrum, we integrate from all wavelengths and for an atmosphere we will have a gray temperature:

$$F^{\downarrow(Thermal)} = \left[1 - 2E_3(J)\right]\delta T_0^4$$

Here is the fact that we have used $\int_0^\alpha B_\lambda(T_0)d\lambda = \delta T_0^4$. As the atmosphere becomes cloudy with respect to thermal radiation, the integral$(J \to \infty)$ tends to zero and a heat fluxδT_0^4 is given. As the atmosphere becomes clearer, the integral tends$J \to \infty$ toward the inclination$E_3(J) \to 0.5$.

The resulting flux is lost because an unabsorbed object cannot emit thermal energy despite its temperature. It is common for Equation 5-18 to be written as follows:

the sky $F^{\downarrow(Thermal)} = \delta T^4$

That the sky temperature is equal to:

$$T_{the\ sky} = \left[1 - 2E_3(\tau)\right]^{\frac{1}{4}} T_\circ$$

Sky temperature is the temperature at which a black body must be brought to that temperature to radiate the same atmospheric flux. It can be seen that the sky temperature is less than or equal to the kinetic temperature of the air T_0. The clearer the atmosphere is than the heat, the lower the sky temperature. When the atmosphere is cloudy.

The temperature of the sky tends towards the temperature of the air movement. Using Equation, we draw the T0 ratio of the sky T in terms of J. Under normal conditions, the sky T is usually less than 10 degrees Kelvin below ambient temperature.

The steady-state temperature of a horizontal surface is determined in part by the temperature of the sky. On a clear, dry night when the sky temperature is low, the ambient temperature of a permanent external surface cools radiantly to a temperature below ambient temperature. This radiative cooling at night is very common in desert areas where the sky is mostly clear and dry. In solar technology, sky temperature plays an important role in the performance of the permanent state of solar thermal panels. We emphasize that ground heat radiation is not really the same, in fact most of the heat radiation received by a vertical surface from the surrounding area Origin. Therefore, the temperature of the sky depends on both the earth and the orientation of the desired surface and the atmosphere itself.

Current Renewable Energy Application in Palestine

* Solar water heating
    * More than 75% of Palestinian residential units use solar water heaters
    * Non residential facilities depend on electricity or fuel for water heating
    * In hospitals, hotels and universities solar water heating covers between 20-40% of the needed amount

(Picow, 2010)

Figure 24. Renewable Energy in Palestine Current and Future

Chapter VII

Direct Conversion of Solar Energy into Work - Photovoltaic Devices

First, the radiation is converted to heat by solar heating devices (e.g., flat collectors and concentrators), and then part of that heat is converted to card, although theoretically all solar energy can be converted to work, but this heat cannot. And some of the heat must be returned to a colder source. The conversion efficiency is limited by the ratio of cold source temperature to hot source temperature.

The lower the ratio, the higher the efficiency may be. In the direct conversion process, we extract the work from solar energy without first converting the electromagnetic radiation into heat. This process does not require hot or cold sources. Although conversion efficiencies may have some practical limitations, they are not limited by the new work formula. To show how a beam of reflection can be converted directly into mechanical energy, we use a device called a vacuum radiometer. The device consists of four blades, one side of which is silver and the other two sides are blackened. The blades are housed inside an empty glass bulb. It has pressure and therefore force on the surface on which it shines. The pressure applied to the silver surface is almost twice the pressure applied to a black surface because they are silver surface and reflect the radiation. A pair of opposite blades is also exposed to a beam of radiation. The force applied to the blade that has a silver surface is exposed to a beam of radiation that is more than the opposite blade whose blackened surface is exposed to radiation. The net torque generated by these forces causes the blades to rotate about a vertical axis.

Consider a parallel beam of radiation with frequency v that shines vertically on the mirror completely flat and completely reflected. The mirror moves at velocity v along the direction of radiation propagation. According to quantum theory, a beam of radiation can be thought of as a stream of particle energy quantum's called photons. The energy and motion of each photon in the radiated light is:

$$E = hv \qquad 6-1$$

$$P = \frac{h}{\pi} = \frac{E}{C} \qquad 6-2$$

In this formula, λ the wavelength and h are Planck constants. If there are N photons in which the flux travels a unit of time, the radiated flux will be as follows:

$$F_{inc} = NhV6 - 3$$

According to the theory of relativity, the radiation that the process reflects has a two-step displacement, so its frequency is:

$$V^1 \left[\frac{1-\beta}{1+\beta}\right] V$$

It is reduced that in this formula C, $\beta = \frac{V}{C}$ is the speed of light. In addition, according to the theory of relativity, it is predicted that the photon flux decreases and reaches the following value:

$$N = \left[\frac{1-\beta}{1+\beta}\right] N$$

Therefore, the reflected flux is equal to: $F_{ref\delta} = NhV'$

Therefore, the net flux transmitted by the irradiated beam is equal to:

$$F_{net} = F_{inc} - F_{ref\delta} = h(NV - N'V') = \left[\frac{4}{1+\beta}\right] F_{inc}$$

Not all of this energy is given to the mirror, but some of it is spent "filling" the space in front of the mirror to find the mechanical power that is really provided. Assume that the mirror moves at a much slower speed ($\beta \leq 1$)than the speed of light. A photon will be repelled by the same amount of motion as P, so the amount of net motion transmitted to the mirror will be 2P. The pressure on the mirror is equal to the product of this. Displacement is the magnitude of the motion of a photon flux, or: 2PN = pressure

Using equations 1-6, 2-6, 3-6 we find that:

$$\text{pressure} = \frac{2F_{inc}}{c} \, (\beta \leq 1)$$

Therefore, the transfer efficiency is equal to:

$$\eta = \frac{\frac{p}{A}}{F_{inc}} = 2\beta = 2\frac{v}{c} \, (\beta \leq 1)$$

Hence the small conversion β efficiency increases linearly with the mirror speed. β It is so small for the heads that the mirror can actually have that no conversion efficiency can be expected. In addition, the maximum pressure exerted on a surface by intense sunlight is less than one-tenth of a billionth of an atmosphere.

Although low radiative pressure may be sufficient to rotate the blades of a vacuum radiometer, this and the conversion of solar energy into friction created that each blade had an area of 9 m2 and a rotational speed of 150 km / h ($\beta = 1.5 \times 10^{-7}$). The force applied to each blade is approximately 10^{-4} kg and its conversion efficiency is less than 0.000003. The important point is that direct conversion is possible, and the other process does not involve a thermodynamic cycle and therefore does not require heat sources to supply or absorb heat. There are processes that convert solar energy directly into chemical energy.

Green plants use the sun's energy to convert water and carbon dioxide into complex hydrocarbons. This is a gradual process, and only a small fraction of the plant's pleasant light is converted into chemical energy. It has been suggested that rummy chemical fuels can be obtained at a cost-effective cost from the living mass, i.e. from the retinas that grow specifically for this purpose. Mass production of alcohol from grain and sugar for vehicle fuel is currently taking place in some countries. One of the most promising methods of direct conversion of electricity generation is the use of photovoltaic devices. Voltage is generated on it. We will examine a common class of these photovoltaic devices, known as PN-bonded devices.

For example, devices made of silicon are capable of producing 0.5 volts per cell in bright sunlight and their normal efficiency. 10 to 12 percent. As we will see in the future, the problem of early technology in p-n photovoltaic devices has been their high manufacturing cost. The theory of photovoltaic devices is relatively complex, so we will only examine the principles of their operation.

Intrinsic semiconductors (pure)

According to the quantum theory of matter, the electrons of a single atom are allowed to exist only in certain states, with a definite energy. It can be attributed to a single energy of more than one state. Of course, in any given case, there can be no more than one electron. When several similar atoms are arranged in a regular set, like a pure crystalline solid, the outer layers tend to expand into energy bands that have several closely spaced states. The states inside these bands are available to the electrons, while the energy levels between these bands are forbidden.

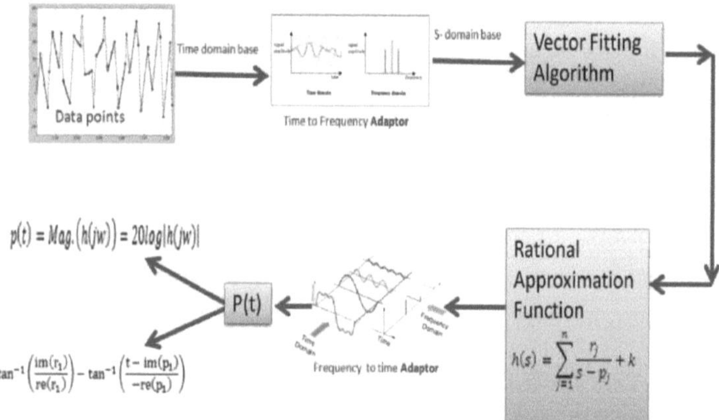

Figure 25. PLOS ONE: Field data-based mathematical modeling by Bode equations and vector fitting

At around absolute zero, the electrons try to occupy the lowest energy states. Electrons are forced to be in layers with increasing energies. Because the number of electrons in each material is limited. Occupied states only reach a certain energy level. If this level is above an energy bar, the solid in question acts as an insulator. In order for an insulator to show electrical conductivity, some electrons must reach a small drift velocity in the vicinity of an electric field. To achieve this speed, electrons must gradually increase their kinetic energy. Of course, this is not possible because a small increase in energy requires that electrons adjacent above the valence band enter the forbidden probe. Adequate valance tape acts as a complete insulator. When the solid temperature rises dramatically, some of the adjacent electrons above the valence band are excited and enter the conduction band. In this way, the material has few electrons that can achieve a drift speed. And therefore shows little electrical conductivity. If the number of electrons in the solid is sufficient to fill a fraction of (for example, copper) the outermost band (conductor). The electrons are defined and move higher into the band. As a result, a relatively large number of electrons achieve the optimum conduction velocity for electrical conductivity, and the solid acts like a metal conductor. A semiconductor at a temperature close to absolute zero is quite like an insulator. Thus, the semiconductor in question has a completely filled valence band and an empty conduction band.

The difference between the two is that in one semiconductor the energy gap between the guide and valance strips are very small. Therefore, at room temperature, a large number of valence electrons are excited by the presence of heat and travel to the conduction band (Figure 2-6b). As the temperature of the material rises, the electrical conductivity of a semiconductor increases sharply. There is electricity in metals and semiconductors. In metals, current occurs entirely due to the movement of free electrons in the conduction band. In the case of semiconductors, the voids left by the electrons stimulated by heat in the valence band are also involved in the electrical conduction process. At the same time, some of the electrons inside the valence band can be excited to go into empty states, thus achieving the desired drift speed for conduction.

The remaining voids in the valence band are said to form positively charged cavities. The electrical conductivity in the valence band can be defined as the current from the holes. When an electric field is applied, the holes produced in the valence band and the electrons produced in the conduction band flow in opposite directions, but the net current is the sum of the currents. It is a hole and an electron. The carrier density for intrinsic or pure semiconductors is denoted by n_i for (negative) electrons and pi for holes (positive). Because in one semiconductor the carriers are led in the form of an electron-hole cavity, pi = n_i. It can be proved that the electron pair concentration of the hole changes according to the following relation with Kelvin temperature.

$$n_i = p_i = AT^{3/2}\ e^{-\in R/2Kt}$$

Where k is the Boltzmann constant and E_g the magnitude of the energy gap is the energy difference between the bottom of the conductor band and the top of the valence band. The constant A is an experimental parameter for a semiconductor $10^{16}cm^{-3}k^{-3/2}$ and its value is generally approx., While Eg is of the order of one electron volt($1eV = 1.6 \times 10^{-19}j$).

$$k = \frac{1}{11600} = 8.625 \times 10^{-5} eV/k$$

According to Equation, the carrier concentration of the T-beam disappears to zero so that the semiconductor gradually acts as an insulator. As T increases, the carrier density as well as the electrical conductivity of both increase dramatically. When an electric field E is applied to a semiconductor, the resulting current density will be equal to:

$$J = \frac{E}{P}$$

Where p is the resistivity. This resistivity can be expressed according to the following formula:

$$p = \frac{1}{en_i(\mu_n + \mu_p)}$$

e The charge of the electron $\mu_p, \mu_n, (1.6 \times 10^{-19}C)$ is the mobility of the electron and the hole, respectively.

When both the carrier concentration and the carrier mobility increase, the specific strength of the material decreases. In fluorine n_i is independent of temperature, the value of pi is zero (there is no cavity flow). μ_n It decreases with temperature. As a result, the resistivity of metals increases with increasing temperature.

In semiconductors, increasing n_i compensates for any decrease in the μ_p, μ_n, and as a result, the resistivity of semiconductors decreases with increasing temperature. In short, it is conceivable that an intrinsic semiconductor has a uniform concentration and a function of the temperature of the electron pair of holes. Equilibrium concentration is maintained through constant heat generation and the consequent recombination of electron pairs in the hole. The resistivity of an intrinsic semiconductor is much smaller than the resistivity of an insulator, but at the same time much higher than the resistivity of a metal. Resistance is very sensitive to temperature changes.

Non-intrinsic semiconductors (impure)

We now consider the effect of adding a small amount of phosphorus to pure silium. A phosphorus atom has a capacity of +5 (one unit more than the capacity of a silicon atom). The net effect of adding phosphorus atoms is reciprocal. It occurs just below the conduction band. Second, the phosphor atoms share electrons to completely fill this surface. Even at room temperature, all of these electrons are excited and move toward the conduction band.

Therefore, a large number of electrons are available for electrical conductivity. In fact, the phosphorus atom of a single electron is shared by conductors, so these atoms are called donor atoms. The density of the electron carrier that is supplied to the donor's surface is:

$$\text{(Article type n)} \quad n \cong N_d$$

Where N_d is the concentration of the atom. It should be remembered that these electrons do not leave a moving hole behind, instead the positive donor ions remain fixed inside the crystal lattice. Although the density of the donor impurity is much lower than the atomic density of the lattice itself, the non-intrinsic carrier is hundreds of times greater than the density of the electron pair in the inner cavity. Which is formed by the thermal excitation of the valence band as a regenerating impurity produces many electrons for conducting action. Minority means that it disrupts the cavities that form in the heat. Reducing the concentration of cavities is achieved by using the law of mass effect:

$$np = n_i p_i = n_i^2$$

$$p = \frac{n_i^2}{n} = \frac{n_i^2}{n_d}$$

Must be established. n_i Equilibrium concentration is an intrinsic mediator. Using Equation 6-7, the concentration of minority carriers is obtained:

$$\text{(Article type n)} \quad P = \frac{A^2 T^3 e^{-\epsilon gkt}}{N_d}$$

Positive donor ions (are constant)

• (Negative (most) electrons (moving)

° Positive (minority) cavities (mobile)

Strip structure in a semiconductor Type (b) Distribution of majority carriers (electrons) Minority carriers (holes) Fixed donor ions.

$$(n = p_i + N_d, n \rangle p_i)$$

Thus, in a n-type substance (e.g., silicon impure with phosphorus), the majority carriers contain a large number of electrons generated by the donor atoms, plus a very small number of electrons produced by thermal excitation. The concentration of the majority carrier is almost independent of temperature and is obtained from Equation 6-9. Minority carriers, on the other hand, are cavities created by thermal stimulation of the valence band. Their concentration varies considerably with temperature. In addition to majority electrons and minority holes, type n material contains a large number of non-mobile donor ions with positive base because the number of electrons is equal to the sum of the number of positive holes and ions. It is electrically neutral.

P-n link

To understand the function of a photovoltaic device, we examine what is formed when a p-n junction is formed. When two p and n substances are connected. The majority carriers in each substance begin to scatter at the graft site to balance their focus. The cavities adjacent to the graft travel from type n to type n, and the electrons begin to scatter in the opposite direction.

As a result, large numbers of carriers reassemble into a thin region called the empty layer, which is basically free of holes and electrons. This action detects immobile ions in each region so that the empty region of type p has a negative charge and the empty region of matter n has a positive charge. As a result, an electrostatic voltage and field are created in the empty layer, which prevents more holes and electrons from leaving the empty layer. The electrostatic potential of n-type material is significantly higher than that of p-type material. This contact potential cannot be measured directly. If the temperature of the p-n junction is different from the temperature behind the electrodes, a thermoelectric voltage will be generated (but here we have ignored this effect).

Transplant photovoltaic devices

Consider a single p-n bond made of p-type silicon wafer with a thin layer of n-type silicon deposited on it. The wafer is called the floor and the deposited layer is called the surface layer. The electrodes are attached to the outer surfaces of the device. The electrode is made of very thin metal deposition for the surface layer. It is also thin, so solar radiation can bond. By irradiating solar energy on a photovoltaic device, some of the photons form an electron-hole pair, and the effect is a stream of light flowing from the n-type p-type material. This light current is:

$$J_P = e^{\circ} n$$

In this connection, n° the type of electron pair generation is a hole (per unit area). There must be a value equal to the light current. That is, there must be a reversible current called the J_j bond current, the value of which is equal to the optical current. This junction current flows from the p-type material to the n-type material and is therefore considered a direct current.

$$J_P = J_j = J_0(e^{V_{oc}/kt} - 1)$$

In this relation, V_{oc} is an open-circuit voltage. A simple model for a photovoltaic cell according to includes a current source that is parallel to a diode. The open-circuit photovoltaic device is electronically specified. It is an isolated device, the photon current produced by solar radiation passes through the diode and is returned by photovoltaics.

As a result, an open circuit voltage is generated on both sides of the diode as well as on both terminals. By solving equation 6-11 with respect to and lattice, we find that:

$$V_{oc} = Kt\lambda n \left[\frac{JP}{J_0} + 1\right] (J_{oc} = 0)$$

The voltage of the open circuit produced by a photovoltaic device with p-n junction depends on temperature, reverse saturation current and photon current. As we will see, the photon current depends to some extent on the intensity of the spectrum of the radiation.

If we short-circuit the terminals of the photovoltaic device. The total photon current is returned through the external circuits and the bond current reaches zero. As a result, as expected, the terminal voltage is zero, so the short-circuit current is equal to:

$$J_{sc} = J_p(V_{sc} = 0)$$

When a charge resistor R is connected to both ends of a photovoltaic device, the fraction of the photon current passes through its parallel diode and the rest of it passes through the said charge. The current passing through the charge is:

$$J = J_p - J_j$$

Solve the above equation for the terminal voltage, we have:

$$J = J_p - J_0(e^{v/kt} - 1)$$

$$V = Kt\lambda n(\frac{JP - J}{J_0} + 1)$$

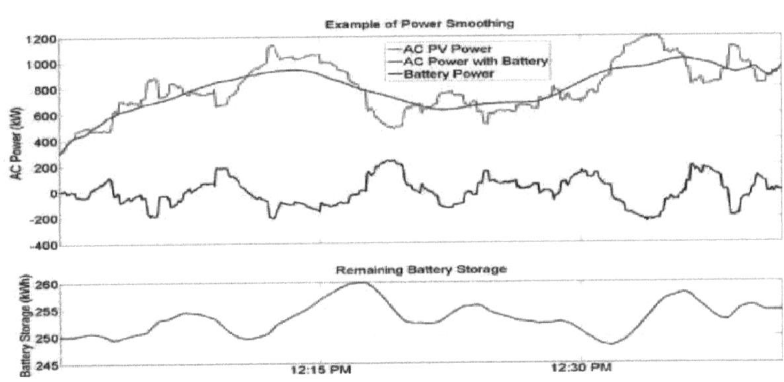

Figure 26. Example of ramp rate control for long time scales

Photon current spectral response

The current characteristic in terms of the voltage of a Rami solar photovoltaic p-n junction can be determined from Equation once the JP, which is related to the radiated flux on the surface layer of the device, is obtained from the equation. Instead, we will present a semi-quantitative experimental analysis. Imagine representing the radiated solar flux F_λ through a spectral distribution. The spectral function represents the radiant energy per second and the unit of surface area per unit of wavelength reflected on the surface layer of the photovoltaic device. Because a wavelength photonλ has energy$E_\lambda = h\upsilon = \frac{hc}{\lambda}$, the number of photons emitted per unit time per unit area per wavelength unit is:

$$N_\lambda = F_\lambda/hc$$

Not all photons produce a photon current. Photons whose energy is less than the energy of a semiconductor band gap cannot even produce a pair of hole electrons. Some photons that have enough energy to produce these pairs are reflected and absorbed by the surface layer before they can do so. they give. All of these factors have been used empirically:

$$n_\lambda = B_\lambda N_\lambda$$

In this regard, the pair production rate is n_λ (per unit area) and B_λ the quantum efficiency is for photons with wavelengths λ. The parameter B_λ for wavelengths greater than the cut-off wavelength λ_0 is zero. The wavelength parameter λ_0 is a photon whose energy is equal to the energy of the band gap. It usually B_λ has the highest value when λit is slightly shorterλ_0 and $\frac{\lambda}{\lambda_0}$ the slope is reduced to zero. Therefore, those photons whose energy is slightly higher than the energy of the band gap are better able to generate photon current. Using the relation:

$$J_{p.\lambda} = e\beta_\lambda N_\lambda = \frac{e\beta_\lambda \lambda}{hc} F_\lambda$$

This relationship can be written in a simpler way

$$J_{P.\lambda} = K_\lambda F_\lambda (6-18)$$

Where the photon current is spectral $J_{P.\lambda}$

And

$$K_\lambda = \frac{e\beta_\lambda \lambda}{hc}$$

Spectral response is a photovoltaic device. The response of a normal silicon photovoltaic device is shown in Figure 6-6. The photon current generated by the whole spectrum is:

$$J_P = \int J_{P.\lambda} d\lambda = \int_0^{\lambda_0} k_\lambda F_\lambda d\lambda = \bar{k} F$$

where in

$$F = \int_0^\infty F_\lambda d\lambda$$

And:

$$\bar{k} \int_0^{\lambda_0} k_\lambda F_\lambda d\lambda / F$$

The definite wavelength is obtained $\lambda_0 = HC/\epsilon_g$ according to the relation. The parameter $\bar{k}$ is common to the mean response and depends on the spectral distribution of the irradiated radiation. Its value is measured in amperes per watt: After substituting Equation, we have Equation:

$$J = \bar{K} F - J_0 \left[e^{\frac{v}{kt}} - 1 \right] \text{ and } V = kt\lambda n \left[\frac{\bar{k}F - J}{J_0} + 1 \right]$$

Using Equation, we obtain the open-circuit short-circuit current:

$$J_{SC} = \bar{K}F \qquad \text{And}$$

$$V_{OC} = Kt\lambda n\left(\frac{\bar{k}F}{J_o} + 1\right)$$

Although the coupling current is linear with respect to the radiated flux, the open circuit voltage varies logarithmically with respect to F. Because the average response depends on the spectral distribution, its value varies with atmospheric conditions and the state of the sun.

To obtain a basis by which to measure the performance of a solar photovoltaic device, we use the spectral distribution of a black body at 5,800 degrees Kelvin. Which operates at T = 300K. These curves are derived from values $\bar{K} = 0.25 mA/mw$ and $J_0 = 5 \times 10^{-10} mA/Cm^2$. For a constant flux level, the output voltage decreases with increasing current drawn relative to its open circuit value. Because the output power density (i.e. power per unit area of the cell) is obtained by the following equation:

$$p_A = V \times J$$

Manufacture of silicon photovoltaic devices

Currently, silicon photovoltaic devices are among the most effective solar cells used. Their construction should start with the purest degree of silicon available (solar degree). This silicon is derived from the more impure silicon (semiconductor grade) used to make electronic components.

The final stage of construction is called regional melting, in which regions of molten silicon move through the mass of matter and carry with them the remaining fine impurities. The purity of the solar degree remaining in the mass of matter is 990999. This pure silicon is kept in the plant by molten radio frequency waves similar to those used in a microwave oven. This causes the material to heat evenly.

The impurity is carefully added to the molten material to obtain n or p type material. The molten material is then crystallized. The crystallization method that is commonly used is the Cheklersky method. A small crystal grain attached to a special clamp is inserted into the molten material. It gradually comes out of the molten material. Its diameter depends on the speed of extraction and grain from the molten material. The lower the speed, the larger the diameter of the blade.

Normal velocities of several centimeters, a fraction of one centimeter per hour, change into beads up to 10 centimeters in diameter. After that, silicon ingots (type p or n) with solar degrees are cut into round wafers with a thickness of approximately 0.075 centimeters. These wafers form a solar cell if the floor material is p. In a thin layer of n-type material, it is formed by spreading the appropriate impurities on the surface of the wafer. The thickness of the surface layer of a photovoltaic device is typically a few microns ($1 = 10^{-4}cm$ microns).

The surface layer is usually coated with an anti-reflective layer. The reflection of radiation decreases during waves to which the cell is very sensitive. When making a cell, the thickness of the floor needle should be so small that the series resistance inside the cell should be reduced and the side should be large enough to be structurally strong. The hole inside the cell should be considerable. Care must be taken to ensure that the metal electrodes are properly connected.

Recent research on thin and amorphous (non-crystalline) semiconductors has shown that photovoltaic devices can be produced at a lower cost. They shine less. If the cost of building batteries can be reduced by a factor of 100 to, say, 40$ per square meter, the total interest can compete even with the usual efficiency of electricity generation, especially since the cost of fossil fuels is rising.

Estimated cost of electricity generation

Currently, the investment cost for SEGS line parabolic solar power plants with a capacity of 30-80 MW is about 2,900 $ to 4,200 $, of which 50 to 70 percent is related to the cost of building and installing a solar farm and 15 to 25 percent is related to the cost of the thermal cycle. And 5 to 10 percent of the initial investment is related to the balance cost of the power plant. Operating and maintenance costs can be divided into two parts: equipment maintenance and personnel costs. For SEGS power plants, the maintenance cost is about 1 per year is 0.9% of the initial investment. The number of personnel required for 30 MW power plants is 33 and for 80 MW power plants is 53 people, the costs of which are 0.9-1.2% of the initial investment of the power plant. Indicates maintenance and fuel for SEGS power plants According to environmental regulations SEGS power plants indicate that SEGS power plants are allowed to supply up to 25% of their thermal energy from fossil fuels according to environmental regulations. However, in our country, due to the cheapness and availability of natural gas, a greater share of the required heat energy can be provided by gas, thereby reducing the cost of electricity production of these power plants determined fossil fuel consumption.

$$LEC = \frac{A * c + O8M + F}{E}$$

In the above relation

C= the total initial investment cost

A= The annual repayment rate

O8M= Operating and maintenance costs per year

F= Cost of annual fuel consumption

E= is the amount of electrical energy produced per year

The annual and subsequent refund rates over the life of the plant (N) and the discount rate (r) are calculated from the following equation:

$$A = \frac{r = (1 + n)^n}{(1 + r)^n - 1}$$

If the life of the power plant is 30 years, the discount rate of 8% and the capacity factor of 30-50% should be considered and 25-50% of the thermal energy should be provided by natural gas. The cost of electricity generation for 30MW power plants is 12-16 C / kwh and 80 MW power plants is 9-11C / kwh

It is about twice the cost of generating electricity from conventional fossil power plants. It is expected that by integrating solar power plants with combined cycle, the cost of electricity generation can be reduced to 7-8 C / kwh. One of the important advantages of these power plants is the reduction of emissions and CO_2 compared to fossil power plants.

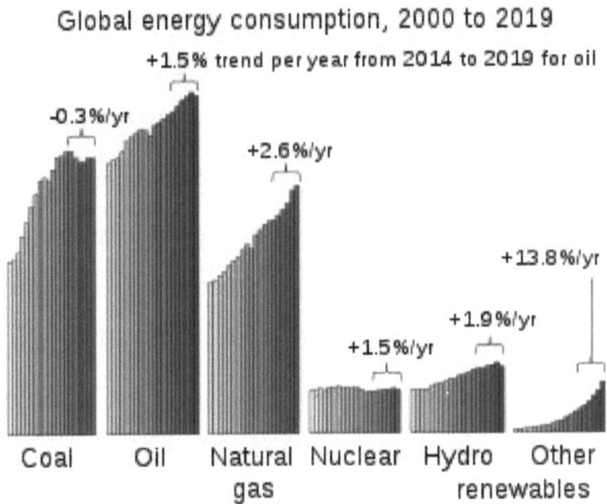

Figure 27. Renewable energy

For example, a sample of 80 Mwe solar power plant with a capacity factor of 35% and the use of 25% auxiliary gas fuel for 30 years to prevent the emission of 310 million tons of CO_2 produced by a similar fossil power plant and de-CO_2 depletion of about 20$ / ton. The operation of this solar power plant will reduce the mentioned cost by 70 $ million. Kanoon-Line Parabolic Solar Power Plant is the first solar power plant that has been used commercially to generate electricity. This is due to the following factors:

➢ Simple technical design of the collector and the availability of their manufacturing technology and mass production

➢ Possibility of using the integrated system with auxiliary fossil fuels in order to ensure the operation of the system in different weather conditions.

➢ High reliability of the system in comparison with other solar thermal power plants due to the proper operating temperature of the collectors.

➢ The collector tracking system is uniaxial and the experiments obtained show that the north-south arrangement compared to the east-west arrangement allows more solar energy to be obtained during the year.

Due to the fact that the concentration part of the parabolic collectors of the linear center is about 100 to 20 and their operating temperature is 300-400 degrees Celsius.

The use of black chromium or ceramic-metal alloys for the absorber surfaces of the receiver and thermal oils as the operating fluid of the collectors is more considered. For capacities less than 30 Mwe, the use of running cycle and auxiliary boiler in the secondary circuit of the power plant is the most suitable design. If the metal is higher than 100 Mwe due to the increase of thermal inertia of the solar field, the use of heater-oil with fossil fuels in the initial circuit of the power plant is more appropriate. The investment cost of these power plants is $ 4200-2900 $/ kwe and the operation and maintenance costs per year are about $ 84-85 $/kwe, and if 20-50% of the thermal energy required by the power plants is provided by fossil fuels, the cost of electricity production is 9-11$ kwh.

References

1. S. Auvil, J. Choe, L.Kellog, 1993, Use of membrane separation to dry gas streams containing water vapor, p. 1-21.

2. L.Kohl, B.Niellsen, 1999, Gas purification, Gulf publishing company, 5 ed., p. 18-210.

3. Association gas processors suppliers, 1999, GPSA Engineering Data Book, 2 ed., 2, p. 1-200.

4. C. Catherine, 2012, Membrane processes in biotechnologies and pharmaceutic, Elsevier, p. 253-293.

5. W. Ronald, 1987, Handbook of separation process technology, Wiley-VCH, p. 35-160.

6. S.P. Nunes, K.V. Peineman, 2001, Membrane technology in the chemical industry, Wiley-VCH, p. 207-300.

7. R.W. Baker, 2004, Membrane technology and application, Membrane Technology and Research, 2 ed., p. 1-70.

8. R.D. Noble, S.A. Stren, 1997, Membrane Separation Technology, principles and application, Elsvier Science, p. 171-189.

9. Z. Gu, W.S. Ho, N.N. Li, 1992, Membrane handbook, Van Nostrand Reinhold, p. 656–700.

10. M. Chakraborty, C. Bhattacharya, S. Datta, 2010, Liquid membranes: principles and applications in chemical separations and wastewater treatment, Elsevier B.V., p. 141-145.

11. Y. Yampolskii, B. Freeman, 2010, Membrane gas separation, John Willey & Songs, p. 8-15.

12. T.A. Vilgis, G. Heinrich, M. Kluppel, 2009, Reinforcement of polymer Nano-composites, Cambridge University, New York, p. 1-5.

13. H. Cong, B. Towler, M. Radosz, Y. Shen, 2007, Polymer–inorganic nanocomposite membranes for gas separation, Sep. Purif. Technol. 90, p. 281–291.

14. X. Ai, X. Hu, 2003, Study on organic–inorganic hybrid membranes, Hu. Jin. 67, p. 654–659.

15.C. Li, H. Shao, S. Zhong, 2003, Preparation technology of organic–inorganic hybrid membrane, Hu. Jin. 18, p. 65–143.

16.R. Stephen, C. Ranganathaiah, S. Varghese, K. Joseph, S. Thomas, 2006, Gastransport through Nano and micro composites of natural rubber and their blends with carboxylate styrene butadiene rubber latex membranes, J. Member. Sci. 101, p. 858–870.

17.M. Smaihi, T. Jermoumi, J. Marignan, R.D. Noble, 1996, Organic–inorganic gas separation membranes: preparation and characterization, J. Member. Sci. 87, p. 211–220.

18.C. Maxwell, 1873, Treatise on electricity and magnetism, Oxford University Press, London, p. 18–123.

19.B.D. Freeman, 2005, Novel nanocomposite membrane structures for H2 separations, Final Technical Progress Report Prepared for the U.S. Department of Energy, p. 1-15.

20.T.C. Merkel, B.D. Freeman, R.J. Spontak, Z. He, I. Pinnau, P. Meakin, A.J. Hill, 2003, Sorption, transport and structural evidence for enhanced free volume in poly(4-methyl-2-pentyne)/fumed silica nanocomposite membranes, Chem. Mater. 15, p. 109–123.

21.V.T. Stannet, 1971, Diffusion in polymers, JohnWiley & Sons, New York, p. 3-8.

22.L. Costello, W.J Koros, 1994, Effect of structure on temperature dependence of gas transport and sorption in a series of polycarbonates, J. Polym. Sci. 32, p. 701–713.

23.H. Cong, X. Hu, M. Radosz, Y. Shen, 2008, Brominated poly (2,6-diphenyl-1,4-phenylene oxide) and its SiO2 nanocomposite membranes for gas separation, Ind. Eng. Chem. Res., p. 1-5.

24.H. Cong, J. Zhang, M. Radosz, Y. Shen, 2010, Carbon nanotube composite membranes of brominated poly (2,6-diphenyl-1,4-phenylene oxide) for gas separation, J. Member. Sci., p. 1-8.

25.T.T. Moore, W.J. Koros, 2005, Non-ideal effects in organic–inorganic materials for gas separation membranes, J. Mol. Struct. 739, p. 87–98.

26.C. Y. Pan, 1983, Gas separation by permeates with high flux in asymmetric membranes, AIChE Journal 29, p. 512-528.

27.W.J. Koros, R.T Chern, 1987, Separation of gaseous mixtures using polymer membranes, Chap.20 in handbook of separation process technology.R.W. Rousseau(ed), John wiley and sons, New York, p. 40-48.

28.S. Rafiq, Z. Man, A. Maulud, N. Muhammad, S. Maitra, 2012, Separation of CO_2 from CH_4 using polysulfone/polyimide silica nanocomposite membranes, Sep. Purif. Technol. 90, p. 162-172.

29.A. Salem,A.A. Ghoreyshi,M.jahanshahi, 2006, A multicomponent transport model for dehydration of organic vapors by zeolite membranes, Desalination, 193,issues 1-3, P. 35-42.

30.G. Dong, V. Chen, 2010, Factors affect defect-free matrimide hollow fiber gas separation performance in natural gas purification, J. Member. Sci, 3, Issues1-2, p. 17-27.

31.L.Haiqing Lin, S.Thompson, A.Lokhandwala, A.Martin, G.Wijmans, D.Amo, B.Ting, C.Merkel, 2012, Dehydration of natural gas using membranes, Part II: sweep/countercurrent design and field test, J. Member. Sci. 432, P. 106-114.

Buy your books fast and straightforward online - at one of world's fastest growing online book stores! Environmentally sound due to Print-on-Demand technologies.

Buy your books online at
www.morebooks.shop

Kaufen Sie Ihre Bücher schnell und unkompliziert online – auf einer der am schnellsten wachsenden Buchhandelsplattformen weltweit! Dank Print-On-Demand umwelt- und ressourcenschonend produziert.

Bücher schneller online kaufen
www.morebooks.shop